D0571921

The Training of Officers

The Training of Officers

From Military Professionalism to Irrelevance

Martin van Creveld

THE FREE PRESS
A Division of Macmillan, Inc.
NEW YORK

Collier Macmillan Publishers
LONDON

The Free Press
A Division of Macmillan, Inc.
866 Third Avenue, New York, N.Y. 10022

Collier Macmillan Canada, Inc.

Printed in the United States of America

printing number

1 2 3 4 5 6 7 8 9 10

Library of Congress Cataloging-in-Publication Data

van Creveld, Martin L.
The training of officers: from military professionalism to irrelevance / by Martin van Creveld.
p. cm.
Includes bibliographical references.
ISBN 0–02–933152–8
1. United States—Armed Forces—Officers—Education. 2. Military education–United States. I. Title.
U408.3.V36 1990
355.3'3'07073—dc20 89–17214
CIP

To my friends and colleagues
in Washington, D.C.

OVER TIME, WHATEVER EXISTS WILL BE COVERED BY SO MANY REASONS THAT ITS ORIGINS IN UNREASON WILL BE OBSCURED.

Friedrich Nietzsche

Contents

Parameters

The purpose of the present study is to offer an overview, as well as a critique, of the way in which commissioned officers are prepared for medium- and senior-command positions. My approach to the question is historical and comparative; that is, I intend to explain how this task has been carried out at various times and places, and also how it has linked up with the military, political, economic, social, and cultural aspects of warfare. The study starts in the earliest times and progresses through the nineteenth and twentieth centuries. It ends with a fairly detailed examination of the present-day American system and some practical recommendations concerning the ways in which that system might be improved.

To put matters into perspective, the first chapter makes a few points about how people are turned into officers, as well as presenting some reflections on what it means to be an officer in modern democratic society. Again the focus is on the American system, and again the approach is a comparative one. In this way I hope to create a clear starting point and set the stage for what follows.

In the United States, the first step that many would-be officers have to take toward the realization of their dream is often that of applying to their congressman for a letter of recommendation. Among modern democratic countries, the idea that admittance to a military academy should depend on such a letter is unique to the United States; it represents a leftover from an earlier, precapitalist military world where promotion depended less on professional skill than on one's social position in the civilian world. Specifically, the system's original purpose was to make sure that the forces, although democratic, would not become too democratic for the country's good. It passed candidates through a sort of old-boy network, thus helping to protect the existing socioeconomic order against those who did not have an adequate stake in it.[1]

In time this aim was lost. Partly because demographic growth meant that members of the governing classes no longer knew each other personally, partly because of populist pressures to democratize the officer corps, what was meant to be an intimate review became a routine bureaucratic matter. Working on the basis of criteria they themselves established, or which were passed on to them by their predecessors, congressmen have delegated responsibility for looking into applicants' credentials to their staffs. Consequently, a situation was created in which power to determine who will *not* be granted entry into some of the most prestigious military academies in the West is currently vested in the hands of a group of young, bright college graduates. Since the U.S. draft was abolished in 1973 most of those who make the selections, whether male or female, no longer expect to see military service or to gain military experience.

Apart from the nominating system, the second outstanding feature of the road toward earning a commission in the United States is that most future officers are designated as such even before they are taken into the forces. This is true regardless of whether they attend a military academy or, as is the case for the majority, participate in an ROTC (Reserve Officers' Training Corps) program attached to a civilian college; in both cases, whatever criteria govern admittance have to be met at the very outset of the hopeful's career. As a result the decision as to who will be allowed to enter officer training rests with examiners and interviewers rather than with people who know the candidate and are familiar with his or her character. Nor, for that matter, does it have anything to do with the candidate's actual performance in military life.

The American system is, in this respect, similar to that of most long-established military powers, including Great Britain, France, and the Soviet Union. However, it is important to recognize that a different system is followed by many of the world's armed forces, including some of the best. Neither the Vietnamese (ex-North Vietnamese) Army nor the Israeli Defense Forces (IDF), to mention but two examples, trust schools to play nearly as large a role in the commissioning system. In those armies most would-be officers are drafted like everybody else. They go through basic training, spend some time in their units, and are then sent through advanced (NCO) training, which counts as the toughest course in the forces. Next they are returned to their units, where they command a tank or gun squad.

The net result of this system is that the officer-candidate examinations, taken, in the Israeli case, after about eighteen months' service, attract an altogether different kind of person. In the United States most officer candidates (except the minority commissioned from the ranks after going through officer training school) are boys and girls just out of high school who select themselves on the basis of personal ambitions, family pressures, and the like. In Israel and Vietnam, on the other hand, they are self-confident young soldiers. Those who take the examinations do so only after having acquired a thorough familiarity with military life, proving their competence by going through advanced training, acting as junior commanders, and earning their superiors' respect.[2]

Third, obtaining a commission in the United States today almost always entails the pursuit of a baccalaureate degree, either at a military academy or at a civilian institution of higher learning; with the result that the U.S. forces have virtually no officers who have not attended some sort of college.[3] Here, too, an international comparison can be enlightening. Long ago the British used to have something like ROTC appended to such prestigious universities as Oxford; but nowadays almost all their cadets, both those with and without a degree, are trained at the Sandhurst Royal Military Academy and in its RAF and Royal Navy equivalents, which do not rate as academic institutions. The present-day Soviet armed forces have a system of degree-awarding military academies adapted from that of the old tsarist Army. With them, however, responsibility for officer training has been decentralized to the individual arms rather than concentrated in the hands of the services.[4]

On the other hand, the Vietnamese and Israeli armies would surely question any system by which aspiring officers receive the bulk of their training in a special institution, separated from the enlisted men whom they will ultimately command. Finally, the postwar West German Bundeswehr has followed the old Wehrmacht and adopted a compromise solution. Although cadets are tested and accepted before enlistment, they receive their basic and advanced training in the battalion and not at an academy. Having thus acquired a thorough understanding of their future subordinates and of military life in general ("operated a machine gun with freezing fingers," as one young captain put it to me), they are sent to officer-training school. It is only after earning a commission that those who have signed on for the full twelve-year career officers' term are sent to attend one of the two so-called *Bundeswehrhochschulen*—which, in most respects, resemble civilian universities.[5]

It is not the purpose of the present volume to criticize the American commissioning system in this respect, nor do I necessarily suggest that officers do not need degrees. Nevertheless, it is legitimate to call the reader's attention to the existence of some scattered evidence that an early college education, with its heavy emphasis on theoretical work and written skill, can actually be harmful to junior commanders whose job, after all, is to lead men in combat.[6]

Fourth, the reader should know just a little about how most military careers in the United States work out. Once a person has received a commission and is not selected out for gross inefficiency, he or she can expect to move up in the military hierarchy to the rank of lieutenant colonel or colonel and to retire while in his or her late forties. A normal officer's career thus spans some three decades; a fact that, in an age of rapid change, itself makes some kind of refresher training not only possible but desirable. This longevity, in turn, is partly the result of the U.S. decision to abolish the draft and establish an officer corps consisting exclusively of long-term volunteers. Therefore, it seems appropriate to remind the reader of the somewhat different course taken by many other countries—including, as it happens, the United States' principal competitor in the international arena, the Soviet Union.[7]

We shall now consider how officers are made and promoted to their position in society. In the democracies, it is taken for granted that an officer should be politically neutral. He or she should be kept, if not actually ignorant of politics, at any rate isolated from them. The officer is not allowed to influence his or her subordinates politically—let alone use political criteria in the exercise of the job. At most, the officer is allowed to tell them that "freedom" is what they are fighting for. However well established this idea may be in the West, it is by no means self-evident. Nor has it been accepted by some of the world's most important armed forces, both past and present.[8]

In the Western countries, an officer's job is limited to offering advice to civilian leaders and carrying out their orders. It is assumed that, precisely because the officer is a "professional," he or she should never presume to decide the most important life-and-death questions facing the state—for example, whether to launch strategic nuclear weapons—let alone to usurp the place of the civilian leaders.[9] Once again, this is not necessarily an assumption that has been shared at all times and in all places.

Finally, this book starts from the premise that commissioned officers do in fact require training and education—in other words, that the various staff and war colleges in the United States and abroad are there to do a real job and not merely to serve as holding pens for vast numbers of unemployed medium-ranking personnel between wars.[10] Nor is my purpose in raising this point merely facetious. Retaining senior personnel in peacetime presented no difficulties in the feudal age, when vassals swore a lifelong oath of fealty to assist their lord whenever called upon and spent the rest of the time on their estates. Nor did the question arise in the age of mercenaries, when officers were dismissed and, if possible, robbed of their pay each time a war was terminated. It first became a serious issue during the eighteenth century. The rise of absolute states, standing armies, and professionalism tended to turn officers into officers and nothing else. As a result, personnel who were to be retained had to be given something to do—and making them study was as often as not simply the cheapest solution to that particular problem. Historically speaking, it was this consideration among others that led to the opening of the first higher institute of military learning in at least one leading country.[11]

In its attempt to examine the way medium- and senior-ranking officers are educated and trained, this study works within the established parameters. It is not primarily concerned with the nature of military professionalism in a modern democratic society, nor with the structure of an officer's career, nor with the way subalterns are made, even though the relevance of these problems to the topic at hand is obvious. Instead, we shall take it for granted that the officer—a fine American lad or lass serving in the regular forces, not in the National Guard—has reached his or her position in the normal way. First he or she showed high potential in a battery of tests, both physical and mental. If he or she was already in active service, the next step led through officer-training school to a commission—a commission, be it noted, which was granted only after the cadet was first formally discharged from the forces and then reenrolled. Alternatively, he or she may have enlisted in a ROTC program or else attended one of the service military academies.

Having graduated from college or the academy by dint of zeal and hard work, he or she is usually sent to a specialist school—armor, say, or intelligence—prior to receiving the first assignment. Once there, he or she is competent on the job. By and large this means commanding the respect of one's subordinates, getting along with one's peers, committing no major blunders, and consequently receiving an outstanding efficiency report from one's superior, which is in turn endorsed by one's superior's superior. Nor does he or she experience any particular difficulty in going through the various short advanced-training courses offered by the service branch to which he or she belongs. With the young officer's foot thus firmly planted on the first rung of the ladder, the question facing the personnel management office in each service is, What next?

Origins

The idea that war is anything but a practical pursuit, hence that senior military commanders—or indeed military commanders of any rank—should prepare for their jobs by undergoing some kind of specialized training and education going beyond practical experience is of comparatively recent origin. In essence, it goes back no further than the eighteenth century; a period that, not accidentally, also witnessed the rise of professional standing armies and thus of an officer corps that had time on its hands between one war and the next.

Describing events that took place around 1000 B.C., the Bible gives no indication that Abner and Joab, acting as commanders in chief for the houses of Saul and David, respectively, received any kind of theoretical training in arms. Joab in particular is portrayed as a rather rough-hewn character who presumably had little use for books.[1] He had started his career as one of the vagabonds who surrounded David during the period he spent as a brigand in the desert.

Such famous classical Athenian commanders as Pericles, Alcibiades, and Nicias, let alone such Spartans as Brasides and Agesilaos, did not reach their position by attending a staff or war college. Plutarch gives us the names of those who taught Pericles philosophy and music.[2] However, he says nothing about any (presumably low-class) drillmasters or teachers-in-arms he may have had, or of the kind of studies that qualified him to lead Athens into the thirty-year Peloponnesian War against Sparta. Of Alcibiades it is said that he was exceedingly good-looking, excelled as a wrestler, and performed his apprenticeship in arms under Socrates, who was in love with him.[3] Nor do the sources tell us much about Nicias, except that he was tall, handsome enough to attract admiration, and the owner of numerous slaves whom he worked in the mines. Apparently those were all the qualifications he needed to become a politician, be elected *strategos,* or general, and be put in charge of the largest single military expedition in the history of Athens—the one that was defeated at Syracuse.

The purpose of the above examples is not to suggest that Greek soldier-statesmen during the classical period were an ignorant lot. On the contrary, it would be difficult to imagine a better-educated set of men than the statesmen Themistocles and Pericles, the historian Thucydides, and the writer Xenophon. Clearly these men had more than mastered the principal cultural skills of their day, including literature, rhetoric, and philosophy.[4] They thoroughly understood the world in which they lived and possessed an articulateness and clarity of thought that has seldom if ever been rivaled. The point is simply that none of them is known to have received military-technical training beyond what was normally provided by the Greek city-state.

Military training in Greece started at puberty. It consisted of physical exercises in the gymnasium, particularly wrestling. Then came weapons drill and learning how to maneuver with others in the close-order formation known as a phalanx. Theoretical preparation consisted of reading Homer, whose role was similar to that of the Bible in a subsequent age. The Greeks were well aware that the weapons and tactics portrayed in the epics belonged to an age antedating the polis and were, as a result, hopelessly old fashioned. Nevertheless, for centuries they served as the principal textbook on the conduct of war as well as on almost everything else. This was because, to most Greeks until the end of the fifth century B.C., war was primarily an exercise in courage and only secondarily an art or science—one reason why it was seldom entrusted to the specialist.

Having reached the age of nineteen, the young men were enrolled as *epheboi,* or young bucks. They spent the next two years roaming the less-inhabited districts, hunting, fighting mock battles, and generally looking for trouble in a process best described as a cross between sport, military training, and an initiation rite. In Sparta, which in this as well as other respects preserved traces of an earlier, more primitive form of social organization, the youngsters also served as a kind of military police, one of their tasks being to stalk and kill Helots who showed signs of causing trouble.[5] This period of training, which in many ways resembled the kind practised until recently by such East African tribes as the Masai, came to an end at the age of twenty-one (in Sparta, thirty). It was then that the *neoi,* or young men, of the polis were solemnly received into full citizenship and given permission to marry.

The Greek citizen-soldier could expect to participate in the campaigns that constituted a normal part of life. If he distinguished himself by reason of birth, wealth, or competence, he could run for political office. The highest of these offices was that of *strategos,* a title that speaks for itself. In any one year in Athens ten *strategoi* were elected, and they divided the work load—whether civil or military—among themselves. If there existed a system to prepare men to occupy these positions, we are ignorant of the fact. Nor would such a system be easy to establish in the face of the notorious reluctance on the part of politicians of all periods to take instruction. Hence we can only assume that senior commanders acquired whatever qualifications they possessed mainly by experience.

During the Hellenistic period, starting around 320 B.C., there was little change. Alexander himself studied with Aristotle. The latter held up Achilles as an ideal and often employed military analogies in his writings; he was, however, a physician rather than a military expert and was certainly not engaged to teach the prince about war.[6] Most of the generals with whom Alexander went on to conquer the world were simply great Macedonian nobles. In the manner of nobles of subsequent ages, they probably underwent their military apprenticeship by serving in related households, gaining experience by campaigning against the neighboring barbarian tribes.[7] Compared with the Greeks whom they conquered, however, their cultural achievements were rather limited.

In this cosmopolitan period of paid, professional armies, a large number of military-technical treatises, dealing with everything from tactical evolutions to semaphore communications, were in circulation. Though such works were presumably read by commanders and other interested persons,[8] there is no indication that studying them constituted a formal requirement for rising in military rank, or that there existed schools for senior commanders. Often the first step to eventual high command taken by an ambitious youth of good family consisted of registering for service in the king's *agema,* or bodyguard. Subsequent appointments would then come his way on the basis of experience and proven success, not seldom mixed with political considerations. There is scattered evidence that commanders sometimes studied tactics with the aid of charts and diagrams, whereas others preferred field exercises.[9] It is with some admiration that Polybius shows us Philopoemen, the great second-century Achaean soldier-statesman, training his own subordinates during country rides—incidentally suggesting both that there was no institutionalized system for doing so and that the task was not considered a self-evident part of a commander's job.

The system by which the Roman republic selected and trained its commanders resembled that of the Greek city-states, except that politicomilitary careers were more highly structured. Of Flamininus, a general *par excellence* who defeated the Macedonian monarchy in 196 B.C., Plutarch says: "From his earliest years he was trained in the arts of war, since at that time Rome was carrying on many great contests and her young men from the very outset were taught by service as soldiers how to command soldiers."[10] A youngster preparing for such a career—almost the only kind, besides agriculture, considered fit for members of the senatorial classes—began by having himself appointed to the staff of some senior commander. At the age of thirty, having already participated in a number of campaigns, he could enter politics and present himself for election to one of the junior magistracies. Most Roman magistracies, but particularly the senior ones such as the praetorship and the consulate, were of a mixed politicomilitary character. Those who held them were *ipso facto* responsible both for civil government and for levying troops and commanding armies in time of war—which in practice meant almost every summer.

As Rome turned from a city-state into a world empire, the number of magistrates in office was often insufficient to command the various forces engaged in military operations in faraway places from Spain to western Asia. This led to the system by which ex-consuls were voted commands, sometimes (like Caesar in Gaul) for years on end. Military experience did not necessarily count for much in these appointments; Cicero, for one, hardly qualified as a soldier, yet as proconsul and governor of Licia he commanded troops and engaged in border warfare. Particularly when the forces in their charge were considerable, these politician-commanders might be reluctant to renounce their positions when their terms ended. Ultimately, the failure of several of them to do so was precisely the factor that led to the fall of the republic.

It was a lesson the emperors took to heart. Whether or not they took the field in person, they always acted as their own commanders in chief and increasingly reserved the term *imperator* to themselves. To fill major commands, they turned to members of their own families, as did Augustus, Tiberius, Vespasian, and the second-century "enlightened" emperors from Trajan to Marcus Aurelius. Men of senatorial rank on their way to the top might start their careers as military tribunes, subsequently moving in and out of political and military appointments. Sometimes the emperors avoided the senatorial class altogether, preferring for such posts as legionary commanders men who were *equites* (knights) and who had no pretensions to govern. Though patronage, seniority, and merit all played a role in the system, their exact relationship is a matter of dispute among historians.[11] Whatever the exact arrangements, they gave rise to a paradoxical situation. Roman soldiers and centurions were tough professionals habituated to war. However, the senior command—prefects and tribunes—often consisted of political appointees, men whose amateur status made them more likely to break into panic.[12]

Beginning in the second century B.C., some of the more studious Roman commanders may have read books. Alternatively, they hired others—often well-educated Greek freedmen, who were becoming increasingly common—to teach them their contents. One need only recall the names of Sallust, Caesar, Livy, and Arrian to realize what excellent military history was written in Rome during the late republic and early empire. Later the histories were joined by substantial military-theoretical treatises such as Frontinus's *Strategematon* in the first century and Vegetius's *De rei militaris* in the fourth. To imagine that senior commanders were required to read this literature in preparation for their jobs, however, is to completely misunderstand what the Romans, and indeed most peoples until the eighteenth century, included under the rubric of war.[13]

During the feudal Middle Ages commanders, regardless of social rank, were normally knights, except when they were women. Though knights began their warlike training as soon as they could get on horseback, that training was almost exclusively practical. The first step in educating an aspiring warrior, or rather a man who was destined for war by reason of birth, was to teach him how to handle a lance while mounted astride a wooden horse. Once he was secure in a real saddle he progressed from one-sided exercises to two-sided ones, and from individual encounters to collective combat. Theoretical training, if that is the correct term, consisted of absorbing chivalric lore. This was done informally, by listening to the tales of one's elders and to the songs of troubadoors, much of which consisted of a paean to war.

Most knights probably first went to war during their late teens, while in the service of some baron in whose household they had been placed as apprentices-in-arms. Another method was for young knights to join together in bands that roamed the countryside in search of a fight against man or beast. *Promotion*—the term itself is an anachronism—depended on catching the eye of some senior baron, either at one of those great tournaments that served as stock exchanges for military talent or else in war itself. Once taken into service, a man would be given a fief to live off of and could expect additional rewards in the form of booty. Appointments to higher posts could be earned on the basis of social position, experience, competence, and proven loyalty to one's lord. Often advancement led through the bed of some heiress, who brought with her into marriage estates and the politicomilitary power that went with them.[14]

A few great princes, such as Richard Lionheart and his rival Philip Augustus, are known to have commissioned special editions of classical textbooks such as Frontinus and Vegetius for their own instruction and that of their sons.[15] During the late Middle Ages men of similar status also constituted a natural market for treatises on chivalric custom written by Jean de Beuil, Honoré Bonnet, and the nun Christine de Pisan. However, the production of books represented a slow and expensive process. Particularly during the early and High Middle Ages, many knights and even great barons were illiterate. Long after this had ceased to be the case, they continued to look down on books as suitable primarily for lower forms of life such as monks, clerks, and burghers. Perhaps more important, war itself was thought of as anything but a learned affair to be mastered in the study. Instead, it tended to be regarded as a vigorous outdoor activity, akin to hunting or sport and in many ways almost as enjoyable.

Thus, the way all these different types of commanders were trained, or rather not trained, for the exercise of high command had everything to do with the political, military, and cultural climate in which they lived. Nor did one attend school, study books, and take examinations in order to become an officer. The concept itself is a modern one, and only entered usage during the late fifteenth century. Its origin is the Italian word *ufficio.* As in present-day English, it could mean either a formal appointment or the physical environment in which the administrative work it entails is carried out.

Toward the end of the Middle Ages, the money economy revived. More and more, kings and princes sought their collaborators not among the great landholding feudal vassals but among officeholders who, being salaried, were easier to control. On the whole, this trend manifested itself *pari passu* in the political and military fields. Before the transformation into standing armies could be completed, however, the stage of mercenary armies had to be gone through. Such armies, which became increasingly common from the time of the Hundred Years War, were commanded by entrepreneurs who received a "commission" from the sovereign. They used the commission in order to raise their own units, relying on their own capital or what they could borrow. Their primary motivation in waging war was, of course, personal profit, often obtained by misappropriating the money entrusted to them for supporting their men.[16]

Though such titles as colonel, captain, and lieutenant (the first was the commander of a column, the second that of a company, the third his stand-in), date from this period, few entrepreneurs of this kind reached the highest ranks. As kings increasingly withdrew from fighting or even commanding in person, those ranks tended to be reserved for great noblemen such as (to mention but three) the Duke of Bourbon serving Charles V, the Duke of Alba representing Philip II of Spain, and the Prince de Condé heading the armies of Louis XIV. Neither the military entrepreneurs nor their aristocratic superiors could be described as professional officers in our sense of the term.[17] The appearance of the latter goes back only as far as the Peace of Westphalia in 1648. By and large, it coincides with the final victory of the absolutist state.

After absolutist government and standing armies had been firmly established, the officer's status underwent a change. Although still for the most part recruited from the nobility and often possessed of independent means, his status as an independent businessman was gradually eroded. Princes of the blood and the scions of grandees could still hope to obtain a commission without doing anything in particular to win it, though as time went on appointments gained by such methods tended to become largely honorary. The majority, however, increasingly found themselves in a situation where commissions had to be earned through study of some kind.

The idea that war consisted of more than the practical art of throat cutting—in other words, that it rested on a substantial body of theoretical knowledge that had to be mastered—is largely a product of modern technology. It first made its influence felt in such fields as artillery, fortification, and engineering as well as navigation. As long ago as the sixteenth century, mathematicians such as the Italian Niccolò Tartaglia and the Dutchman Simon Stevin were offering instruction in these subjects to young persons contemplating a military career. During the seventeenth century prominent Spanish, French, and Dutch commanders set up private military academies. The original purpose of these institutions was to relieve commanders of the burden of training the youngsters who applied for a post at their headquarters; subsequently they were regularized and brought under government control. Finally, during the eighteenth century, these courses developed into schools for subalterns, including the military ones at Paris, St. Petersburg, Munich, Wiener Neustadt, and Woolwich, and the naval ones at Dartmouth, Le Havre, Toulon, and Brest.[18] Nor was this movement toward formal education restricted to the military domain. Rather, it was part of a much wider phenomenon associated with the Enlightenment, which resulted in the opening of schools for many kinds of specialties ranging from mining to civil engineering.

The role which these institutions played in military life during the days of the *ancien régime* should not be exaggerated. Many of them—including the first, the École Royale Militaire in Paris—were intended more as indoor relief for the sons of the impoverished aristocracy than as places for serious military learning. Members of the nobility who could afford to follow the wars at their own expense continued to receive commissions on the strength of favor, experience, or monetary payment. Promotion from the ranks, though far from common, was also possible, in particular during wartime. Either way, the principle that no one could receive a commission without first attending a school of some kind only became established during the nineteenth century. Thus, though Napoleon himself was a product of the academy at Brienne, his own army became notorious for putting a field marshal's baton in every soldier's knapsack. Nevertheless, the academies' very existence shows that military training and education beyond the most elementary level had finally come of age.

The academies opened their doors to youths in their early teens and sent them out when they were about sixteen years old. The faculties were made up mostly of officers, often elderly personnel invalided out of the forces, with a few civilians thrown in. The military part of the curriculum consisted mainly of weapons drill and tactical evolutions; the latter were often taught with the aid of models (this period saw the rise of the tin-soldier industry) and war games played by moving wooden blocks on a map. Much the most important nonmilitary subject was mathematics, which in the age of linear tactics and magazines was regarded as the true cornerstone of war. Others were history—this being a period when diplomatic and military history were almost equivalent to history as such—geography, and foreign languages in various combinations. Weapons drill and minor tactics aside, the only subjects taught at all the academies everywhere were fencing and dancing, both of which had long been established as indispensable to anyone with aspirations to be a gentleman.

Perhaps more important from our point of view, the military academies helped create a demand for texts. This demand, in turn, could be satisfied thanks to the invention of printing in 1453, which led to the rise of a large military literature, first in Greek and Latin and then in Italian, Spanish, German, English, and, above all, French. Men whose social position destined them for a military career, such as Prince Maurice of Nassau or King Gustavus Adolphus of Sweden, are known to have studied this literature as part of their general education. Others did so as the occasion presented itself, for example, the imperial commander Raimondo Montecucculi—the son of an Italian family that was half noble, half bandit—during the four years he spent as a prisoner in Swedish hands.[19]

The eighteenth century, with its emphasis on enlightenment in every field of human endeavor, saw a massive outpouring of military-theoretical and military-historical volumes. Some were authored by the most outstanding generals of the age, from Maurice de Saxe and Frederick the Great down. Others were written by famous staff officers such as Pierre Bourcet, whereas others still were the product of amateurs eager to try their hand, such as Guibert. These books, in turn, were reviewed and discussed in a periodical literature that tended to specialize as time went on. Toward the very end of the period discussed in this chapter there even appeared the first manual for staff officers—a clear sign that the age of higher military education had finally arrived.[20]

The officer corps of the various countries were thus thrown open to talent, or at any rate the talent of those who could show some kind of social pedigree. Meanwhile, senior command continued to be exercised, for the most part, by members of the high nobility—even to the point where anyone who was promoted to *maréchal de France* by that very fact became a cousin of the king. However good or bad the qualifications they possessed—and often they were impressive—these noble commanders found themselves overwhelmed by the demands of contemporary war.

The years between 1550 and 1700 witnessed the so-called military revolution. Helped by the printing press, improved maps, and better communications, governments were able to raise armies on a scale not seen in Europe since the fall of the Roman Empire. Supplying these forces and equipping them, managing and administering them, presented formidable tasks. The difficulty was all the greater because, by tradition, many of those problems had scarcely been regarded as part of war at all. Commanders cast in the heroic mold did not, as a rule, consider them part of their job; they were low-prestige occupations, fit only for civilian contractors, secretaries, and aides de camp.

Casting about for better-qualified personnel to coordinate these functions, commanders from the middle of the eighteenth century concentrated on the staff of an officer known as the quartermaster or quartermaster general. Originally the quartermaster was a minor figure responsible for riding ahead of an army on campaign and staking out the camp every night—hence the derivation of the term *staff*. To carry out his duties he had to know some mathematics, and the same applied to his assistants. The location of this group in front of a marching force made them well suited for the gathering of intelligence, both geographical and military, and for this purpose too some education was required.

The combination of intelligence, a good military education, and a forward position caused the quartermaster's duties to grow and grow until he became the quartermaster general. The point came where the office clashed with and gradually grew to eclipse the traditional military cabinet, charged with the commander's paper-work. By the middle of the eighteenth century a good quartermaster who had the confidence of his commander was in a position (to paraphrase Frederick the Great) to collaborate on every aspect of war except operational planning—and here, too, his proximity to the center of action would enable him to learn all he needed to know.[21]

As organized from the last decades of the seventeenth century on, quartermaster staffs were small outfits numbering between five and seven officers. Consisting of personnel that did not form part of the regiments, they were set up separately for each war and campaign and disbanded as soon as those were over. At the end of the Seven Years War, however, this situation changed. The first to have the idea of retaining "staff" officers and making them study in peacetime was, as far as can be established, Frederick the Great. His académie des nobles, an extension of an earlier system under which twelve promising young officers were attached to the king's suite each year, opened its doors in 1763. Its graduates were designated to act as assistants to commanders of large formations; given competent performance, they themselves could expect to rise to high rank.[22] The académie was not, however, a great success. Too much emphasis was put on abstract science, mathematics in particular, and not enough on practical military affairs. Thus, credit for setting up what was to become the forerunner of the first true staff college must go to Philippe de Ségur, who took over as French minister of war in 1780.

In 1783, after the end of the American Revolutionary War, Ségur feared that the army was about to lose a number of experienced officers. He accordingly offered peacetime employment to a total of sixty-eight men, most of whom held ranks from lieutenant to major. Some he used on his own personal staff. The rest were assigned to a newly established section of the ministry, where their task was to study such professional military subjects as history, geography, reconnaissance, and practical science. Though the staff college never established a monopoly over the selection and training of staff officers (let alone senior commanders, a purpose for which it was in any case much too small), the new institution was an immediate success. When the War of the First Coalition broke out in 1792, those who had passed through its ranks included the man who really won the Battle of Marengo, Desaix; the future field marshals Kléber, Soult, Ney, and Gouvion-Saint-Cyr; and Napoleon's own chief of staff, Berthier.[23]

To conclude this brief historical introduction, the idea that officers need any kind of specifically military training and education beyond what practical experience can provide does not, for the most part, antedate the sixteenth century. When it finally made its appearance—coinciding, not accidentally, with the modern concept of the officer itself—it was at first limited to the technical services, from which it spread in all directions. Officer training as such dates to the establishment of absolutism in the seventeenth century; most modern military academies date to the eighteenth. It should be noted that this was several hundred years after the university education of other professionals—such as theologians, lawyers, and doctors—had become a matter of course. Even so, the principal function of the new academies was to cater to those who could not afford anything else. Such great commanders as Raimondo Montecucculi, John Churchill, Maurice de Saxe, and Ferdinand of Brunswick never passed through their doors. What education they received was acquired at home, ending at an age between twelve and seventeen.

The earliest formal institutions offering anything like an advanced military training to personnel who had already won their commissions were founded toward the end of the eighteenth century. In the main their appearance was due to two factors. First, war was growing in scale and becoming more complicated, leading to the incipient rise of what were subsequently to become the first modern general staffs and causing people to reflect on the training that their personnel required. Second and equally important, if the regular armies of the time were to retain the services of highly qualified professional personnel in peacetime, these people had to be given something to do. Certainly the schools opened by Frederick the Great and Ségur were not intended to teach any subjects except for the military conduct of war. The contents of the instruction that they did give will form the main subject of the next chapter.

Comparisons

The training and education given to prepare officers for command before the beginning of the nineteenth century were always the products of the political-military-social circumstances prevailing at each time and place. Specifically, during most of history before 1800, senior commanders obtained their appointments less because of the subjects they had studied or the schools they had attended than on the basis of the qualities they supposedly possessed. As demonstrated by the endless catalogs of talents required in a general—catalogs that dot military literature from its beginnings to the eighteenth century[1]—what normally counted was less knowledge than character.

From Homeric Greece through the Middle Ages and down to the end of the *ancien régime,* character in turn was often supposed to be the result of breeding—in other words, genetic descent—and social position. Breeding was joined by whatever went under the name of *education,* which according to the tastes of the governing classes might include anything from hunting to music. Such advantages, real or presumed, made a man fit for exercising political office. Very often this also meant military office, if indeed the sequence was not reversed.

Thus, to ask why senior commanders before 1800 did not receive what is today regarded as an advanced military education is itself to misunderstand the nature of their position. Moreover, it reflects another error that is as serious and perhaps more common. Today we are accustomed to divide war into strategy and tactics. Though the terms themselves date to ancient Greece, our modern use of them dates back only as far as the second half of the eighteenth century.[2] During most of history before that time, the conduct of war (strategy) and of battle (tactics) were virtually the same. By this I mean that, apart from battle and sieges, the conduct of war was scarcely regarded as a military problem at all.

There were many reasons for this, the most important ones probably being the limited range of the weapons in use and the tendency, governed by the lack of telecommunications, to operate in single massive blocks rather than in a dispersed manner as was later to become the case. An enemy more than a few hundred yards away might, as far as his capacity to inflict injury was concerned, as well be on the moon. An army on campaign normally spent no more than a handful of days in actual combat. Much the largest part of the season was always taken up by something best described as a mixture of tourism and large-scale robbery.

These facts, of course, were reflected in the way commanders operated. To an Alexander, a Richard Lionheart, a Gustavus Adolphus, a Frederick the Great, to say nothing of their principal subordinates, the meaning *par excellence* of military command was not to sit in some office, study maps, receive messages, issue orders, monitor operations, plan strategy, and make an occasional visit to the front. Rather, its significance was to command the army in battle—even if this did not include, as often continued to be the case down to the end of the seventeenth century, actually fighting hand to hand.[3] With war defined in such a way, it is scant wonder that the *military* education of those whose station in life designated them for high command started with physical exercises and proceeded through weapons drill. Hardly ever did it extend very far beyond grand tactics—that is, how to deploy and employ forces in battle. Though a literature on these subjects did exist and was sometimes consulted by commanders, then as now they were best mastered not in the classroom but in the field.

Tactics aside, everything else that a commander had to know fell into one of two overlapping categories. There were, on the one hand, the auxiliary sciences; these might include castramentation (the art of setting up camps), supply, transport, administration, health care, signaling, and so on—in short, everything necessary to enable an army to exist on a day-to-day basis or else engage in specialized operations such as sieges. Since senior command was usually the reward for long practice, most commanders probably acquired sufficient experience to know at least something about these subjects. Often they were entrusted to specialists, who either were members of the army or else followed the commander in his capacity as a private individual.

The remaining types of knowledge necessary to the conduct of war—such as rhetoric, astronomy (before portable clocks were invented, useful for determining the time), geography, and history—did not fall under the rubric of military education at all. There were no specifically military rhetoric, astronomy, geography, and history, only history, geography, astronomy, and rhetoric in general. There was no military astrology, only Astrology with a capital A—employed, among other things, to forecast the outcomes of engagements and to determine lucky days on which to fight. Such fields, too, were the province of specialists attached to headquarters. The idea that they formed an important part of war proper, and should therefore be studied by would-be commanders and their assistants as part of their military education, only arose late in the eighteenth century. Not accidentally, that was also the period when the first modern staff colleges were founded.

In short, the explanation for the lack of any institutionalized form of advanced military education before the eighteenth century is to be found in two fundamental factors, both of which had everything to do with the general social, political, and military background. First there was the way senior commanders were appointed, which is practically same as saying that the word *commander* itself carried a different meaning. Second, there was the way military knowledge, and behind it war itself, was structured; so little did reading and writing count as part of war that they were entrusted to a civilian *chef de cabinet*. Thus, when advanced officer training and education finally did become institutionalized, that fact reflected a revolution that was taking place in men's minds as to what war is and what its conduct involves.

Our starting point for study will be the Prussian Kriegsakademie, an institution of advanced military learning, which during much of the period from 1815 to 1945 was regarded as the best of its kind and served as a model for many of the rest. We shall then advance country by country, until we end up with the USA.

Prussia-Germany

In Prussia advanced officer education originated, during the reign of Frederick the Great. Around the time of the outbreak of the Seven Years War, the Army's quartermaster general, von Schmettau, commented on the generals' inadequate education; the king's alleged response was that, by the time they had risen to be generals, they were often beyond help.[4] Still, he took three steps to remedy the situation. First, he provided five schools—at Wesel, Magdeburg, Breslau, Königsberg, and Berlin—where officers could study during the winter months from November to February. Second, he encouraged the establishment of military libraries in garrison towns. Third, he began appointing promising young officers for tours of duty on his own staff, in the hope that they would learn something. The last-named procedure was at first limited to wartime, but later it was made permanent by turning the Berlin school into the académie des nobles.

Nor was the king satisfied with institutional arrangements. An indefatigable worker, he wrote extensively on military affairs. Over a period of almost fifty years he poured forth a steady stream of military histories, letters of instruction, and so-called political testaments. Attempting to offer subordinates food for thought, he even composed *Kriegsgesänge,* or war songs.

Though these beginnings were not unimportant, modern advanced military training really got its start in 1801. It was in that year that Gerhard von Scharnhorst, an officer in the Hanoverian Army who had already made something of a name for himself as a staff officer and a military writer, transferred to the Prussian Service.[5] When he took over as director of the Militärakademie (as the académie was now known, following a burst of Francophobia), he found a decrepit institution urgently requiring reform. At that time those enrolled in the academy were still young officers of noble descent, selected on the basis of no formal examination system. The permanent faculty consisted of just two men. One was a seventy-year-old major of the engineers, a veteran of the Seven Years War who lectured on military geography, fortification, siege work, and field encampments. The other was Johann Gottfried Kiesewetter, an influential popularizer of Kant, whose task was to teach mathematics and logic. The rest of the instruction was provided by visiting lecturers. Among them were civilian professors as well as senior officials working for the Prussian war department.

Scharnhorst's first move was to make the Militärakademie independent of the Berlin school for officers of which it had previously formed a part. Next he appointed a third permanent faculty member, whose function was to teach applied mathematics; to Scharnhorst this represented a way of making sure that pure mathematics would receive even greater attention. Later, arguing against the Enlightenment view, which held that the science of war could be deducted from first principles, he added military history. The study of foreign languages, particularly French and Russian, was revived and remained an important part of the curriculum right until the time the Academy was finally dissolved in 1945. The student body was expanded to forty officers. As a means for raising the academy's prestige, King Frederick Wilhelm III was persuaded to allow some of the meetings to take place in the royal palace of Charlottenburg. Among Scharnhorst's students during this period were several future chiefs of staff and, of course, the young Clausewitz.[6]

Organized on these lines, the academy, like the rest of the Prussian Army and state, was thoroughly shaken up by the defeat suffered at Jena in October 1806. A period of turmoil followed, and what had previously been an isolated debating society turned into a hotbed of military reform. In 1810 Scharnhorst carried out yet another reorganization. Renamed the Allgemeine Kriegsschule (General War Academy), the school now offered a nine-month program. The curriculum included mathematics, tactics, strategy, staff work, weapons science, military geography, German language, foreign languages, physics, chemistry, horse care and mess administration in various combinations for officers belonging to the different arms.[7] Instruction consisted of lectures, seminars, and workshops. Unlike that employed at ordinary officer training schools, it put a heavy emphasis on free study and independent work by the students.

By this time, the commanders of the old Frederican army had been thoroughly discredited. The same applied to the system under which promotion to higher rank was governed by descent and patronage. Consequently, the academy was in a position where it could insist on the need for theoretical study and professionalism as essential qualifications, if not for holding high command, at any rate for serving on the staffs of large formations from division upward. The rise in its status enabled the academy to change the system by which promising young officers were selected for study. Instead of trusting to a combination of recommendations by superiors and noble descent, the academy instituted written entrance examinations. Even though taking the examinations was voluntary, and even though in later years only about 20 percent of those who tried passed, their very existence acted to press an officer corps that was often recalcitrant into the direction of study.

After the fall of Napoleon, with "peace reigning hard," officers accustomed to the hectic pace of military operations found themselves with time on their hands. This permitted the course to be extended and its duration to be set at three years. When the young Moltke, then a lieutenant who had just transferred from the Danish service, attended the academy between 1823 and 1826, the curriculum for the first year consisted of mathematics, topography, general history, statistics, artillery, tactics, sketching (theoretical and applied), French, and horse care. The second year was devoted to more mathematics, some mechanics, military geography, fortifications, German literature, strategy, more sketching, French, and natural science; whereas the third was spent studying "the history of selected campaigns" (mostly between 1704 and 1815), siege warfare, general literature, the history of the general staff, and still more sketching. At this time, and until the outbreak of World War I, the program of study was intended to make the students proceed gradually from general theoretical subjects, such as mathematics and history, toward specifically military ones such as military history and fortifications.

By 1912 the curriculum was as follows:[8]

First Year	*Hours per Week*
tactics	4
weaponry	4
fortifications	1
naval warfare	2
military sanitation	1
military justice	1

general history	1
geography	3
mathematics (or French, English, Russian, or Japanese)	6
Second Year	
tactics and general staff work	6
military history	4
fortifications and fortress warfare	2
sketching (soon to be rendered obsolete by photography)	1
transport	2
general history	2
political science, administration	3
law, finance	1
mathematics (or French, English, Russian, or Japanese)	6
Third Year	
tactics and general staff work	6
military history	4
general history	3
English, Russian, or Japanese	4

As is evident from this table, tactics constituted the most important single subject studied. Spread out over the three-year course, in one form or another it took up one-third of the curriculum; tactics; moreover, also formed the principal subject taught during so-called *Geländeübungen* (terrain exercises) and the annual *Stabsreise* or staff rides. Under the influence of the Kriegsakademie tactical problem-solving became a minor hobby in Prussia. There appeared a small library of commercial volumes containing *Aufgaben* (problems) and *Lösungen* (solutions), much like modern crossword-puzzle books. Just as present-day military periodicals often carry equipment-recognition quizzes, so every issue of the *Militärwochenblatt* contained at least one tactical problem from the 1850s on.

Moreover, it is important to understand what the Germans meant by *Taktik.* Present-day English uses the term to describe the operations of minor units in immediate contact with the enemy. In German military parlance before 1918 it meant something more like "the operational art of war"—that is, maneuvering and fighting large formations up to and including corps level. Hence the academy's product was not battalion commanders—in fact, most battalion and even regimental commanders before 1945 never passed through its halls. Rather, it was majors—later, lieutenant colonels—qualified to perform as chief of staff *cum* operations officer (Ia) of a division and, eventually, higher formations.

Still, what set the Kriegsakademie apart from similar institutions in other countries was less its curriculum than the exalted status that it enjoyed both in and out of the military. In part, this status reflected the glory of the general staff itself; that staff had been transformed by the victories over Denmark, Austria, and France from an unimportant department in the ministry of war into "the most honored institution in the world" (Moltke). To a large extent, however, this was due to the deliberate efforts of General Eduard von Peucker in his capacity as Prussian chief of military training during the 1860s.

Fusing the ideals of Alexander and Wilhelm von Humboldt with those of Scharnhorst and Clausewitz, Peucker really wanted to ensure that the Akademie would not be just a professional school but a university, the seat of higher military learning that would originate new ideas and spread them throughout the army. It was to open its doors "to a group of exceptionally talented, professionally committed officers of all arms who were already in possession of an adequate military-scientific education and wished to extend it." It offered them both "a first class, in depth, professional education" and "higher formal instruction in those sciences which should be regarded either as the foundations of military art or as auxiliaries thereto."[9]

To explain the success of this ideal, it is important to recall the social structure of Bismarck's Germany. Traditionally, Prussia had been dominated by the nobility. In 1866, following a historic meeting of the Chamber of Deputies, the rapidly rising commercial-industrial middle class formally entered an alliance with the state. What united both groups was a fervent admiration for everything military—with the emperors presenting themselves as the first soldiers of the Reich and an appointment as reserve officer a must for anyone wishing to be someone. In addition, the German middle classes more than their counterparts in other countries had long been excluded from political power. To compensate, they often followed the example of Kant and put a heavy emphasis on *Bildung,* or education, as they still do.

This combination had the fortunate effect of putting the Kriegsakademie in a unique position. Almost alone among all the institutions in the Reich, and to an extent unequaled in any other country, it could command the reverence stemming from both the military and the intellectual traditions. Consequently its prestige stood far higher than that of any mere civilian university, and it was able to attract the *crème de la crème.*[10]

To enter the academy, an officer had to pass a series of written, anonymous examinations whose topic was set annually by the chief of the general staff and which were prepared and supervised by the faculty. Military subjects included *Waffenlehre,* tactics up to reinforced-brigade level, fieldcraft, and combat-engineer service. Other subjects were foreign languages, history, and geography. All these were studied at the *Abitur,* or high school level, which was the approximate equivalent of the degrees awarded by all but the best modern American colleges. Preparation for the examinations usually required four to five months and was supposed to be carried out by the officers in their own free time. During the last years before World War I some eight hundred officers entered the competition every year. Of these, perhaps 20 percent passed the test and were accepted as students.[11]

During Scharnhorst's lifetime, the officers entering the Militärakademie were for the most part senior lieutenants in their mid-twenties. However, the century of almost uninterrupted peace after 1815 had caused promotion to slow down, with the result that the average age of entrants rose into the early thirties during the last decades before 1914. The academy's graduates—there were no formal examinations, but the students were graded by the staff on the basis of their written work and personal acquaintance—gained the equivalent of ten years' seniority *vis-à-vis* their comrades. Their first assignment was to serve for six months in an arm other than their own. Then they were called to Berlin and taken into the general staff on a probationary basis. Having served for a year, and having received a suitable efficiency report, they were formally taken into the *corps d'élite* and allowed to wear the coveted double crimson stripes on their trousers.

A very important aspect of life at the Kriegsakademie, and one that probably did as much as any other factor to ensure its success, was the institutional affiliation of the faculty. Though other countries were to follow the Prussian example in setting up a general staff corps, nowhere was membership in that corps so exclusive or its prestige so high as in the German army. Not only were the permanent military instructors members of the corps, but the academy was able to attract men on their way to the top. Thus the future field marshals Colmar von der Goltz, Hindenburg, and Ludendorff—the last named to the faculty at his own request—all served a tour of duty teaching before progressing to other duties. While there is no reason to believe that the quality of their instruction was better (or worse) than that of their comrades in other countries, such men could clearly serve as role models to their students, who looked up to them and hoped to be in their shoes one day. Thus, service on the faculty was a coveted assignment and an honor. Far from sidetracking a man's career, it carried substantial financial benefits and was regarded very much as an essential step on the way to a senior post.

The Kriegsakademie, like the general staff itself, was formally abolished in 1919 under the terms of the Treaty of Versailles. However, in practice both institutions continued to exist. Though left-wing propaganda undermined the social position of the army during the first years after the war, it soon recovered. With fewer than 4,000 officers when it was established, the Reichswehr was really able to select. Guided by von Seeckt in his capacity as chief of the *Truppenamt,* as the camouflaged general staff was now known, select it did. The principal instrument remained the written examination. It was renamed *Wehrkreisprüfung* (after the *Wehrkreise,* or military districts, into which Germany was divided) and made obligatory for all officers. Those who passed, normally just over 10 percent of the total, were called to attend a so-called *Führergehilfelehrgang* (commanders' assistants' course). Though promotion in the Reichswehr was generally slow, a system of preferred advancement caused the students' average age to drop slightly. Most now entered the academy while in their late twenties—probably the best age at which to learn or study.

Held in the provinces in order to escape the prying eyes of the Allied Armistice Commission in the years following World War I, the *Führergehilfe* courses were the equivalent of the first two years' study at the Kriegsakademie. The curriculum became even more varied than before. It included, in one combination or another, tactics and military history (the two most important subjects by far), military organization, *Waffenlehre,* air-to-ground cooperation, air defense, general staff work, foreign armies, logistics, transport, and counterintelligence. The principal nonmilitary topics still consisted of foreign languages, international relations, and economics.[12]

The method of instruction was predominantly practical. There was heavy emphasis on independent written work by the students, who were given problems (*Aufgaben*) and had to hand in solutions that were then criticized by the faculty. Often they had to work under deadline pressure, a method that supposedly reproduced the stress of battle to some extent. There was, however, no specialization. Officers of every branch—operations, intelligence, supply—continued to study together, which may help explain why German interarm cooperation during World War II very often gives the impression of a smooth, well-integrated team. Cross-familiarization with the various arms was carried out by means of mutual visits, each lasting several weeks. The overall goal was to produce a body of military experts thoroughly familiar with every aspect of their profession and capable of taking over from each other at a moment's notice.

As the second year of training drew to an end, approximately one-third of the students were selected for the final year of study, which took place in Berlin. Subjects continued as before, but the curriculum now included a heavy sprinkling of civilian experts—often the leading specialists in their fields—who, while personifying the ideal of *Bildung,* were only too happy to be invited. Their lectures embraced political, international, and economic affairs. However, the core of the Berlin faculty continued to consist of senior, experienced, promotable general staff officers who could serve as role models for their students. Handpicked by the chief of training of the general staff, their quality is perhaps best brought out by listing the names of some of them: They included the future field marshals von Kluge, List, Model, and Paulus, as well as the future colonel-generals Adam, Guderian, Halder, Jodl, and Reinhardt.[13]

By and large, the system Seeckt had established for the Reichswehr was retained by the subsequent Wehrmacht. In 1933, after the incipient expansion of the army and the greatly increased demand for staff officers, the number of study years was cut to two; however, this stage came to an end as early as 1935, when the Kriegsakademie was formally reopened with an impressive ceremony attended by Hitler. The system of obligatory entrance examinations remained in force, acting as a spur for study. So did the curriculum, with its heavy emphasis on tactics, military history, and foreign languages. At the insistence of the chief of the general staff, Ludwig von Beck, the meaning of *Taktik* was narrowed down. From 1935 on it excluded operations above divisional level. The change may have been motivated by the desire to focus on immediate requirements during a period of breakneck military expansion. Nevertheless, it was regarded by some critics as an error.

Students continued to be graded by their instructors on the basis of no fixed system—the establishment of formal examinations was discussed but abandoned. Aside from the numerous written exercises, great emphasis was put on personal acquaintance between instructors and students—an acquaintance promoted by communal evenings, field trips, and the *Schlussreise* (final trip) held each year. Faithful to Clausewitz's dictum that war is a matter of character above all, the evaluators looked for such untranslatable attributes as *Anstaendigkeit* (uprightness, decency, and reliability), *Verantwortungsfreude* (joy in responsibility), *Seelenkraft* (spiritual and mental force), and the ability to work long hours under pressure without sacrificing quality. The system by which successful graduates were taken onto the general staff for a period of probation also remained in force. Last but not least, instructors continued to be selected from among highly qualified, experienced, general staff officers who could serve as role models for their students and who themselves could look forward to a military future.

In 1938 scandal rocked the Wehrmacht as the minister of war, general von Blomberg, was accused of having married a prostitute. He was forced to resign, and the post was taken over by Hitler. A reorganization of the armed forces followed. The navy and air force—the latter commanded by the formidable Göring—tried to seize the opportunity in order to assert their equality *vis-à-vis* the army, which was larger, older, and more prestigious. One outcome of this struggle was the establishment of a Wehrmachtakademie for all three armed forces, which in practice amounted to a downgrading of the old Kriegsakademie. Although it was commanded by an army officer, General Adam, the new institution was staunchly resisted by the army and was consequently not a success. Only a single class ever went through its doors, and after the outbreak of World War II it languished.

The Kriegsakademie owed its foundation to Scharnhorst and to the incipient consciousness that war was not just a practical art but a science that could be subjected to historical and analytical study. Like that of its parent organization—the great general staff—

its social position was greatly reinforced by the victorious wars of the mid-nineteenth century, and it was indeed to this position that its success was in large part due. The system of selection, the three years' practical and theoretical training, and the probationary period ensured that the product—a lieutenant colonel capable of acting as divisional chief of staff—would be thoroughly trained and competent. It also ensured that the army's key officers, those on the staffs of major formations and those appointed to the general staff in Berlin, would share a common outlook and a common language.

The thorough training, however, constituted only one aspect of the Kriegsakademie's contribution. Equally as important were the intimate mutual acquaintance and esprit de corps it fostered among a relatively small body of elite personnel, destined to occupy senior positions throughout the army. Uniformity of thought, in turn, enabled the army to give individual commanders a large measure of independence. It thus served as the basis for the decentralized command system known as *Auftragstaktik* (mission-type orders), which was introduced by Moltke and which constituted a key element in the army's success. Still, the Akademie was never able to exercise a monopoly over officers designated for high command. Though most generals after 1870 passed through its doors, promotion on the basis of proven competence in the *Truppendienst* (line service) remained possible. Insofar as this system acted as a spur to the academy, its effects were probably beneficial. It also ensured that officers whose talents were not primarily of an intellectual kind—Rommel being the best-known example—would not be disqualified for that reason alone.

During World War II the pressure of events caused the length of general staff training to be cut. It varied from six months to a year, and was finally fixed at eighteen months in 1943. Two steps were taken to compensate for this. First, in the academy itself there was some tendency toward specialization. Second, short courses were organized for newly appointed senior commanders (division and up) who did not have the benefit of full academy training.[14] These measures, as well as the retention of the probationary period on the way to entering the general staff, and the fact that many senior posts were occupied by graduates of the old Akademie, prevented a serious deterioration in the quality of command. The Wehrmacht's expertise on what was now known as the operational level of war—from brigade to army inclusive—remained second to none.

Though the Wehrmachtakademie never amounted to much, the lack of an interservice school did not present a real problem in World War II. Given Germany's geographical position, the wars conducted by its army and navy were bound to be largely separate anyhow. Where interservice cooperation was required, as for example during the invasions of Norway and Crete in 1940 and 1941 respectively, its quality was generally superb. The German method in such cases was to put an air force officer in overall command. Since most senior Luftwaffe commanders from Field Marshal Kesselring down were themselves ex-army officers who had graduated from the Kriegsakademie, mutual understanding was easy to achieve.

On the negative side, it would be an error to see in the academy anything but a school for staff officers and future commanders of large formations. Courses dealing with politics, international relations, and economics were included in the curriculum; however, their importance was not comparable to that of the strictly military subjects or even to that of foreign languages. Nor did the academy even attempt to provide instruction in the technical and scientific subjects so vital to modern war.

Toward the end of the Weimar period, an attempt was made to correct these deficiencies by instituting the so-called Reinhardt Courses and by sending officers to attend courses at the University of Berlin.[15] The program was designed to acquaint a small number of exceptionally gifted officers with the nonmilitary aspects of war, including politics, industry, finance, transport, and media, as well as the history of Germany's most important rivals in the international arena; all this without losing sight of the fact that the task of officers is to command in war and not to act as experts on a smorgasbord of nonmilitary subjects.

By and large, these efforts were to remain stillborn. The rapid expansion of the armed forces in the thirties created a shortage of staff officers. This fact, and the high value put on things martial as opposed to almost any other human activity, precluded any thorough instruction in the nonmilitary aspects of war. Nor was the movement toward a broader type of military education assisted by the antiintellectual attitude typical of the Nazis. Increasingly during the last years before World War II, the academy's product was a military technician, pure and simple.

Partly as a result of these factors, German military excellence in the field during World War II made itself felt mainly up to and inclusive of army level. It is a remarkable fact that the famous names that emerged from the war—the Balcks, Guderians, Kesselrings, Kluges, Mansteins, Mellenthins, Models, Rommels, Students, to name but a few—were, every one, corps and army commanders. All were tough fighters, excellent tacticians, and masters of the operational art—briefly, the German equivalents to Patton. It is impossible to imagine any of them playing the role of a Marshall, an Eisenhower, or a MacArthur, all of whom were not just generals but diplomats, politicians, and finally statesmen. The one exception to this rule was Field Marshal Gerd von Rundstedt. He, incidentally, belonged to an older generation and had to be recalled from retirement in 1939.

The fact that senior German officers were never required seriously to study the nonmilitary aspects of war may also help to explain another phenomenon, namely their inability to stand up to Hitler. The astonishing domination by the "Bohemian corporal" of the commanders of the world's proudest army was the product of many factors. Among them were his unequaled prestige as head of state and, during his early years in power at least, as Germany's savior. Also there was the terror inspired by the Gestapo and SS, whose roles tended to grow as time went on.

Still, one should not overlook the fact that Hitler could claim, as he often did, that he knew more than his generals, and that this claim was not entirely without foundation. Having seen four years' service at the front in World War I, Hitler possessed a worm's-eye view of the fighting that was unequaled among many of the staff officers in his headquarters. Though an autodidact of rather disorganized habits, he was also a student of economics, history, politics, and many other subjects.[16] During the early years this often enabled him to outargue his generals by shifting the debate to fields with which they felt unfamiliar.[17] Later, as the tide turned, it enabled him to buck them up. Confronted with an officer intent on telling him how bad things were at the front, Hitler would talk about new miracle weapons that would save the situation, or else about the last-minute reversal in the fortunes of Frederick the Great.

Thus, our final verdict on the old Kriegsakademie must be that, as a school for staff officers and experts on the operational art of war, it was unrivaled in its time and has probably remained so to the present day. However, it did not do enough to prepare men for the exercise of command at those elevated levels where military, political, economic, and social affairs merge and become one. This shortcoming in turn was due partly to the way in which the German army and German society defined war and partly to the fact that the prestige of officers was so high that they could scarcely be bothered to master any topics outside their own profession. In the end, this was deliberately exploited by Hitler. Viewed in this way, the Kriegsakademie must be held coresponsible for the German army's success, but also for its eventual failure.

France

While Berlin during most of the eighteenth century was a backward provincial town, the position of Paris as the world's intellectual center was undisputed. It was in France that the experimental spirit, first fostered by Francis Bacon around 1600, found its most fertile soil. It was in France, too, that the *encyclopédistes* attempted to put all human affairs on a new scientific footing, incidentally devoting 800 out of 17,000 entries to military affairs. France was the first country to have a school for civil engineering, the École des Ponts et Chaussées. It was also the first to be fully mapped by the new method of triangulation, which marked an enormous advance in cartography. Finally, it was in late-eighteenth-century France that the first schools of higher learning that were not universities—namely the École Normale d'Administration and the École Polytechnique, were established. True, neither of these was, properly speaking, a military academy. However, the subjects taught in all these academies were of obvious relevance to the ability of the military to carry out its function, and this fact was recognized right from the beginning.

The move toward the institutionalization of higher military education, which got under way in the last years of the *ancien régime,* was interrupted by the revolution. The incessant wars and the slogan of *les carrières ouvertes aux talents* (careers open to talents) caused tens of thousands of officers to gain vast practical experience, while the importance of theoretical study was downgraded. Napoleon himself was a keen student of his profession and also had a good general education, but he tended to look down on the administrators and technicians produced by the contemporary equivalent of a staff college. Among the marshalate, some had passed through Ségur's establishment before 1789. Most of its members had only graduated from officer school, however, whereas a few—Massena being the best-known case—did not even do that.

At the lower end of the scale, the revolution not only opened the officer corps to the sons of the middle classes but wrought a vast change in the social position of the military. After order was restored and the Consulate was established, many of France's best young minds entered the army. However, for them, too, study did not present the highway to promotion. Instead, everything depended on catching the eye of some marshal or, better still, that of the emperor himself. The way to start a promising career was to get oneself attached to a major headquarters as one of those *officiers d'ordonance* who crisscrossed Europe gathering information and carrying messages.[18] The result was gross favoritism and nepotism as high-ranking officers awarded positions to members of their own families or to their friends' minions. The situation was saved only by Napoleon's insistence that staff officers serve tours of duty in the line as a condition for being promoted.

Though the reasons behind them changed, by and large these antiintellectual tendencies maintained themselves after 1815. The restored Bourbon monarchy distrusted the educated middle classes because of their revolutionary potential. In turn, the bourgeoisie found its energies fully absorbed by banking, commerce, and rapid industrialization. Each side for its own reasons supported the return to a professional army, which naturally led to the revival of an aristocratic officer corps. Typical of the French service, however, was the fact that the corps also included a sprinkling of ex-NCOs, often originating in the lowest social classes. They had proved themselves on active service and subsequently rose to senior command positions.[19]

Of the two categories comprising the new officer corps, neither was particularly inclined toward school learning and study. The members of the aristocracy, following the feudal tradition, considered it beneath their dignity. The tough, hard-bitten ex-NCO types were for the most part incapable of engaging in it. Thus it happened that the France of the Restoration at first had nothing like the old Prussian académie militaire, let alone the Allgemeine Kriegschule established by Scharnhorst.

In 1817, the question of setting up a general staff corps and a school to prepare officers for serving in it came up for consideration by the minister of war, Gouvion-Saint-Cyr, who had been one of Napoleon's marshals. He referred it to Baron Paul Thiebault, who seventeen years previously had distinguished himself by publishing the first general-staff manual in any language. The baron put forward a proposal for a two-year course, to be entered each year by twenty-five to thirty lieutenants of two years' service on the basis of competitive examinations.

Thiebault's proposals were approved. In 1818 there was set up an École d'Application d'État Major, whose seat was in Paris. The annual intake was supposed to consist of young second lieutenants fresh out of officer-training school and, later, of one-third *polytechniciens* supplied by what was fast becoming France's most prestigious institute of higher learning. However, right from the beginning the school experienced trouble in attracting suitable candidates, with the result that there were sometimes only fifteen or twenty students in a class. One reason for this was that the school was not exclusive; instead of standing at the apex of the military education hierarchy, as was the case in Prussia, it was merely the coequal of the artillery and engineering academies and drew on the same manpower pool. Another was that the students' expenses were not paid, nor were their salaries sufficient to allow them to live in Europe's most expensive capital.

The two-year course included geography, statistics, topography, cartography, military reconnaissance, history and administration, artillery-science, and siege warfare—the last subject being taught with the aid of Vauban, whose century-old work also served as a fundamental text in every foreign academy.[20] Thus, it is notable that the French paid relatively much less attention to tactics than did the Prussians. On the other hand, there was a much heavier emphasis on general theoretical studies. This reached the point that, in 1857, a member of the Académie Française complained of the school's pretensions to rule in matters of astronomical and physical science.[21]

The preferred method of study consisted of long hours of classroom work, relieved by hikes during which the students were instructed in mapmaking. Though the demands made by the program were impressive on paper, the school's overall intellectual level was mediocre. Its commandant was a *maréchal de camp d'état major,* his deputy a lieutenant colonel. Apart from the teacher of

German, a civilian, the entire faculty was military. They carried the ranks of major (battalion commander) and captain—though it is important to remember that the prestige associated with these ranks a century and a half ago was much greater than it is today. Some of the instructors were too young, others too old and intent only on keeping a sinecure that gave them a comfortable life in the capital. The students, it is said, summed up the lessons by commenting that "everything depends on topographical and atmospheric circumstances."[22]

To correct the graduates' lack of practical experience, Saint-Cyr decreed that they should continue their careers by serving first in a cavalry and then in an infantry regiment. Actual admission into the general staff depended on passing an examination, described as "rather tough" by Thiebault.[23] As in Prussia, those taken into the staff formed a separate body. They had their own promotion procedures and their own distinctive uniforms; at a time when the rest of the army was ordered to grow whiskers (blond ones had to be covered with black shoe polish), general staff members were forbidden from doing so. They were known as the *corps royal* and were never able to get themselves accepted by the troops.

Once established on these lines, the École d'Application underwent comparatively little change after 1820. In 1844 there was a proposal to open the school to officers who were not Saint-Cyriens—to those, in other words, who had been promoted from the ranks and possessed no educational qualifications. Another group of reformers sought to bypass the school entirely, suggesting that the general staff be opened to officers who had not passed through its halls. Though both attempts were rejected, they indicate the amount of hostility provoked in the army by the existence of a *corps d'élite* with intellectual pretensions. Nor were the school's opponents entirely unsuccessful in their attempts to reduce its importance. The graduates' period of service in line units was stretched and stretched. It finally reached twelve years instead of the original three, thus giving members plenty of time to forget whatever they might have learned. Promotion within the corps was excruciatingly slow, a man taking as long as eighteen or twenty years to reach the rank of captain. Consequently, when the history of the Corps d'État Major was written in 1881, it turned out that not one officer of the classes 1856–1860 had yet succeeded in making colonel.

The way the École itself was organized, as well as the fate that befell its graduates, help explain why it never gained the kind of prestige associated with the Kriegsakademie; and also why not a single line officer ever took advantage of the opportunity to enter the general staff corps by simply taking the examination. There were, however, additional reasons. Between 1815 and 1864, the Prussian Army saw very little action—all it did was to mobilize in 1830, 1851, and 1859, and to engage in some skirmishing against the revolutionaries in Baden in 1848–49. *Faute de mieux,* they were compelled to study. Meanwhile the French gained vast operational experience by fighting in Greece, Spain, Algeria, the Crimea, Italy, and Mexico. One result of this was to reduce the prestige of the École d'Application and of book learning generally. Since promotion was easier to gain in the field than by service on the staff, the French senior commanders in 1870 were, almost to a man, veterans who had proved their competence in recent conflicts.[24] Among the army and corps commanders, the only ones who had been to the École were Trochu and Macmahon—not that this constitutes a recommendation. These commanders seemed much tougher and better prepared than the intellectuals on the other side, headed by the seventy-year-old "decrepit Moltke" (as the French press called him), who had never commanded anything larger than a battalion, and that in peacetime.

Those officers who did graduate from the school and enter the general staff corps fell into four groups. Some were employed at the *dépôt de guerre,* where their duties were mostly of a technical nature—preparing maps and inventories, assembling statistical data on the resources of foreign countries, and the like. A second group was sent to serve as military attachés at embassies in foreign countries, whereas a third worked in the interior helping to run training camps, depots, lines of communication, and so on. The fourth group, which should have comprised the general staff proper, was not organized on any systematic basis. France had nothing to parallel the *Grosser Generalstab* in Berlin, nor did it even have a *Truppengeneralstab* with clearly defined duties. Instead, the graduates of the École were put at the disposal of general officers in command of corps, divisions, and brigades and entrusted with any task that might come to hand.

The outcome of this haphazard system was that, when Napoleon III went to war in Italy in 1859, the highest-ranking officers on his own staff—one major general, one divisional general, and two brigadier generals—were not graduates of the school.[25] Of those who were, none had attained a rank higher than colonel, and the majority were mere captains. Among major subordinate units, some had general staff officers serving on their headquarters whereas others did not. A list of which posts should be occupied by such officers, and what their duties were, did not exist. Thus the French Army did not make systematic use of what intellectual talent it had at its disposal. As contemporaries such as the ex-military attaché in Berlin pointed out, this represented perhaps the greatest single shortcoming of its organization.[26]

After the downfall of the Second Empire, the Republic (which succeeded it) proceeded to abolish the professional army by degrees. When universal conscription was finally established in 1889, it brought into the army the well-educated sons of the bourgeoisie who had previously avoided service by hiring a substitute. Side by side with this went the usual intensive search for the causes of defeat. Different parts of the army put the blame on each other, until everybody finally agreed that the Battle of Sedan had been lost by the French schoolmaster and won by the Prussian one.

As part of the educational reforms that followed, the old École d'Application was closed down in 1876. In its place was set up the École Supérieure de Guerre, a different and much better institution. Its first head was General Lewal, a well-known military historian with expert knowledge of the Napoleonic period in particular. During the years before 1914, members of the faculty included Grouard and Colin, both of them notable military historians whose works are still worth reading; Bonnal (later promoted to general and appointed commandant), likewise a military historian of note; General Debeney, who was to become chief of staff after the war; the future field marshal Henri Pétain, notable for the impassive irony with which he lectured; and the future field marshal Ferdinand Foch, who by way of preparing himself for the job wrote a four-hundred-page book entitled *La conduite de la guerre.* Clearly, an assignment to the École no longer amounted to either military exile or a sinecure. In the words of one general who was in a position to know, the faculty represented "whatever was best and most distinguished in the French Army of the day."[27]

Admission into the École, now set at eighty a year, continued to be by means of a competitive examination. Applicants tended to be captains aged around thirty; they included graduates from

France's most prestigious schools, Saint-Cyr and the Polytechnique, though the latter were notorious for confining their interest to technical matters and refusing to pay attention to the higher aspects of war. Students were graded on entry and again when they left the school.

Under the direction of General Lewal, the two-year curriculum was restructured with an eye toward practice. It thus came to resemble the German one more closely but probably went too far in eliminating purely scientific subjects and those that were not strictly military: No room was left for fields such as economics or politics. Instruction was limited to strategy, general tactics, the tactics of the various arms, fortifications, general staff work, administration, naval warfare, mobilization, railway transport, military health service, and German (Russian was optional). This being the heyday of positivist, "scientific" military history, much attention was devoted to the wars of the period 1793–1871 in an effort to discover and apply what Foch called "the eternal verities" of war. Lectures and written assignments alternated with the very important visits to exercise grounds, during which the students were given the opportunity to participate in the development of new weapons and tactics. The course also included extensive trips to historical battlefields and reconnaissance along the frontiers.

As was also to be the case at the U.S. Army War College, an attempt was made to entrust faculty and students together with the development of the army's doctrine. This resulted in such regulations as the *Réglement de manoeuvre d'infanterie* of 1875 and the *Instruction sur le combat* of 1887, but led to a situation in which the work of the École duplicated that of the general staff. While the task of writing regulations was taken away from it, however, the school's standing remained high. It became an informal center from which military doctrine—including the erroneous one of *élan*—and the preeminence of the offensive that was to cost the French so dear in World War I was spread in the army and beyond.

The enormous losses suffered in World War I, and possibly a general malaise in French society, caused the prestige of the armed forces to fall after 1918. The prevailing intellectual climate was strongly antimilitary, and the army found itself isolated in the nation. The outcome was a drop in the percentage of those who joined the officer corps from the great schools such as the Polytechnique, whereas the number of ex-NCO types rose sharply.[28] The intellectual development of the army and of the École de Guerre were arrested. A large surplus of experienced officers at the top of the military hierarchy meant that the promotion of their juniors—in other words, École graduates—was slowed down. The opening of a Centre des Hautes Études Tactiques and of a Centre d'Études Tactiques Interarmes signified, as the establishment of similar institutions within the military often does, that study was becoming a substitute for exercising command instead of a stepping-stone towards it. Apart from Colonel Juin, later to acquire fame at Monte Cassino, the instructors were no longer the exceptionally strong personalities they had been before 1914.

The two-year course remained in force. Strategy was dropped from the curriculum and transferred to the Centre des Hautes Études, which in practice meant that few officers ever studied it. Its place was taken by a variety of nonmilitary subjects, reflecting—in this case—both a realization that war had changed and a decline in the prestige of the military profession as such. In the early thirties the military part of the program was modified to take account of advancing technology. In addition to the old cycle of infantry, cavalry, and infantry tactics, the students were now made to visit the Centre d'Instruction des Chars de Combat at Versailles, where they received some instruction in the technology, organization, and use of tanks.[29] To the extent that there was an accepted doctrine, it consisted of rehashing the lessons learned in World War I and particularly the Battle of Montdidier in 1918. "Fire kills"—Pétain's old dictum—had now been turned into official dogma, so that any future attack would have to be slow, deliberate, and preceded by a massive artillery bombardment.

Though much attention continued to be paid to military history, the old self-confidence with which a Foch or a Maillard had expostulated on it was gone. Instead of being regarded as a source of positive doctrine, history was now seen as a method by which "students, using their good sense, character and personality" could develop their decision-making powers.[30] Practical work is said to have consisted of drawing up exceedingly detailed, pedantic orders in writing. These were apt to be returned with the comment that they constituted examples only—neither the sole possible solution nor one that applied to all similar situations.[31]

In 1936, a Collège des Hautes Études de la Défense Nationale was set up by the minister of defense, Édouard Daladier, in consultation with the eighty-year-old Pétain. Its purpose was to offer twenty officers and seventeen civilians from the various ministries a four-month course in the higher conduct of war, including civil-military relations, economics, finance, the military potential of France and the other principal powers, and foreign policy. As happened in Germany also, the college was not welcomed by the services; in this case it was the navy and air force that resented what they saw as an attempt to subordinate them to a single, unified command that would be dominated by the army. During the two years that the college existed, it apparently paid much attention to questions of economic warfare and little to anything else. For this it was criticized, but attempts to make it into a real center for the advanced study of war came too late.[32]

Thus the history of the French Army's institutes of higher military learning from 1815 to 1939 is more chequered than most. It reflects continual tension between colonial and continental warfare, theory and practice, pro- and anti-intellectual movements in the army, and various types of officers. Whereas the Prussian-German Army after 1811 or so never had any doubt concerning the need to select a small number of officers and give them the highest possible military instruction as a virtual *sine qua non* for eventual promotion to high command, the French for long periods were none too clear on this point. Before defeat opened their eyes, the Corps d'État Major was a relatively unimportant, neglected part of the army. It did not serve as a vehicle for selection, nor were its members used to the best advantage when the test came.

Between 1871 and 1914 a dramatic change took place. When the École Supérieure opened, it acted as an elite institute, with attendance mandatory for anyone with aspirations to rise to senior rank. A first-class faculty was put in charge, with the result that the quality of the instruction offered in Paris was probably as good as anything at the Berlin Kriegsakademie. The École's problems, on the other hand, stemmed from excessive confidence, which, however, was indispensable if France was ever to overcome a growing demographic shortfall *vis-à-vis* its principal prospective enemies. The outcome was the doctrine of *élan* with all its disastrous consequences.

After World War I the pendulum swung in the other direction. Except for a vague hope to repeat the successful battles fought against a disintegrating German opponent late in 1918, the French

Army no longer *had* any doctrine. The establishment of the Centre des Hautes Études and later of the Collège des Hautes Études had the effect of splitting the École Supérieure, which could no longer act as *the* most important locus of advanced military training and doctrine. Instead it became merely one of several schools and research centers whose main function was to act as holding pools for a surplus of old, experienced (and, to add to the problem, victorious) officers. The inability of the École to attract first-rate faculty, and hence to offer first-rate instruction, may also have reflected a decline in the prestige of the armed forces in general as well as a change in the prevailing sociointellectual climate. Together, these factors played a large part in the debacle of 1940.

Britain

Compared to the Prussians, the French, and even the Russians, all of whom started giving their officers an academy training during the first half of the eighteenth century, the British were latecomers. It was only in 1799 that the Duke of York, acting in his capacity as commander in chief, caused the Royal Military College to be opened at High Wycombe, near London. Before then the only military academy in England had been Woolwich. Woolwich, however, was a technical school specializing in artillery. Youngsters aspiring to a military career in either the infantry or the cavalry were first subjected to scrutiny by the adjutant general's office. If deemed to be of acceptable gentlemanly birth and bearing—the forms used did not have a heading for education—all that was necessary for them to be commissioned into the regiment of their choice was that they should pay a few hundred pounds.

When reform came, it was due mainly to the efforts of Colonel John Gaspard le Marchant, an ambitious officer whose ultimate fate was to be killed leading a cavalry charge at Salamanca in 1812. Having gained firsthand experience of the disastrous deficiencies in staff work that accompanied British operations in the Low Countries in 1793, Marchant was able to enlist powerful support for his schemes. A sum of 10,000 pounds sterling, which Prime Minister Pitt received in return for the land to be used by the new school, may have helped smooth the way. Like the Frederican Académie des Nobles, Marchant's school consisted of a junior and a senior department. The former was to evolve into the Sandhurst Royal Military Academy, the latter into the Camberley Staff College. Only much later was the connection between the two institutions formally dissolved.

A document written in 1800 outlined the functions of the senior department, incidentally shedding an interesting light on what the early nineteenth century understood as the nature of war and staff work:

> The instruction shall primarily consist in explaining the nature of the country and forming the eye to the judicious choice of positions and the conduct of offensive and defensive war. It shall comprise the manner of reconnoitring a country, so far as concerns military operations; of laying down a plan of it with expedition, of making choice upon the ground of the situation for camps, of marking out and considering of field works and intrenchments which are to defend a given position; together with the methods of encamping, of quartering camps and of marching armies. Those theories shall be enforced, and exemplified by different applications of them, made upon the ground by positions taken and by imaginary marches from one camp to another. . . . All orders relative to those different duties shall be given and explained in the same manner, as would be required on actual service.[33]

As in the case of contemporary institutes in other countries, the senior department was intended less for preparing a military elite than for training a small group of specialists for duty on the quarter-master general's staff. Attendance was, accordingly, purely voluntary. The department tended to attract young officers without family or fortune who had no other hope for advancement except, they hoped, their brains. The handful of instructors were a mixed lot, consisting mainly of civilians; the army at that time did not have officers capable of serving suitable theoretical instruction.[34] When the instructor in tactics, a French émigré general by the name of Jarry, retired in 1806, an effort was made to secure the services of Scharnhorst to head the college; a telling comment both on the reputation of the man himself and on the still international character of the military profession at the time.

Admission to the college was by a superior's recommendation followed by an oral examination. The two-year program put a heavy emphasis on tactics, which to Jarry meant the paradelike evolutions of Frederick the Great; his lectures on the subject earned him the nickname of "Old Pivot." Also included were those usual mainstays of military science—geography, sketching, and mathematics—as well as foreign languages (French or German). Though the department was much too small to make more than a marginal impression on the army as a whole—it never had more than thirty-four officers in residence at any one time, and normally the number was considerably smaller—it remained active throughout the Napoleonic Wars. It helped train such officers as Sirs Charles and William Napier, John Hope, and George Murray, all of whom were to win lasting fame either while serving under Wellington or in India.

This auspicious beginning was not destined to be carried far into the nineteenth century. The restoration of peace after 1815 caused the large units of the British Army to be dissolved, thus obviating the demand for staff officers. As an economy measure, government financial support was withdrawn. The college was made to support itself from the tuition and fees paid by the students (under the government of Margaret Thatcher there is some talk of returning to that system). The length of the course was reduced from two years to one. When Marchant's successor as commandant, General Howard Douglas, left in 1824 no one was appointed to take his place.

By default, the task of leading the senior department fell to John Narrien, a civilian professor of mathematics. He was the author of textbooks with such titles as *Euclid, Conic Sections,* and *Practical Astronomy,* which he inflicted on his students. Under his rule the passing-out examination, first instituted in 1808, was turned into a farce and finally abolished altogether. Drained of its military instructors, the senior department did not offer courses in staff work, strategy, or military history. Instead it focused on theoretical science, allegedly equated with "humbug" by the diminishing number of officers who bothered to attend.[35]

Economy apart, other factors were responsible for the department's decline. The Duke of Wellington, acting first as secretary of war and later as prime minister, had received his own military "graduate education" commanding troops in India; he did not see the need for a training he himself did not have. To him, a good officer was less a studious pen pusher than one who had gained practical experience and could hold his liquor with his comrades in the regiment. Even more deleterious to any attempt at serious military study was the fact that Britain, alone among the major European powers, still relied on the system whereby officers had to purchase their promotions. This put a premium on wealth, connections, and successful marriages as ways to attaining senior rank.

A turning point in the history of the senior department, as of so much else in the British forces, came as a result of the Crimean War. Seen from a military point of view "the war that refused to boil" was uninteresting. However, it was the first time that the course of operations was reported by telegraph. In the newspapers, sob stories took the place of strategy. A large reading public was made aware of the fate of troops at Sebastopol. Their sufferings were widely believed to be the outcome of shortcomings in signals, logistics, organization, auxiliary services, and a competent staff organization to coordinate the lot.

A War Office investigation into the condition of the senior department, held in 1855, revealed a lamentable state of affairs. There were only two professors. One of those was Narrien, by this time blind with age, who for better or worse also taught topography, fortification, and artillery science. An English-language book on staff work did not exist, nor did an authoritative text on tactics. The number of those who applied to take the entrance examination had dropped to fifteen per year—no wonder, since students at the department were required to pay tuition, room, and board to the tune of thirty pounds sterling. In addition, in typical British fashion, it was revealed that out of 216 who had graduated since 1836, only 7 held staff appointments in 1854 and only 15 in 1856. Conversely, out of 291 acting staff officers on the army list, 15 had attended the senior course.[36]

The War Office investigation was followed by a parliamentary inquiry. Its most valuable witness was the veteran General Howard Douglas. He recommended that the student intake be expanded from fifteen to thirty, that the curriculum be reformed to put greater emphasis on fieldwork and on military history, and that appointments to staff posts be made absolutely dependent on graduating from the course.

In 1856, just as the war came to an end, the Duke of Cambridge entered what was to become his forty-year tenure as commander in chief. He and the secretary of war, Lord Panmure, fought each other as to who would rule over military education, a dispute Queen Victoria finally resolved in the duke's favor. With the Prince Consort's assistance most of General Douglas's recommendations were pushed through. The association between the senior department and the Royal Military Academy was dissolved, and the Camberley Staff College assumed an independent existence at the Italianate country house it still occupies. A Council of Military Education was set up that laid down guidelines concerning qualifications to be demanded of entrants and prepared a curriculum.

When the council finally published its General Order and Regulations in December 1857, the result was, theoretically at least, a model of its kind. From 1860 on, graduating from the college was to be made into a prerequisite for service on the staff. The number of students was to be doubled: out of an annual intake of thirty, twenty-five were to come from the line, cavalry and guards, whereas five were to be detailed from the ordnance corps. The number of faculty was increased to a respectable nine, four of whom taught military subjects. The two-year curriculum was restored. It now included military history and the strategy of past wars as well as terrestrial navigation, surveying, sketching, and pure mathematics. Study was to be followed by officers serving two periods of six months each in an arm other than their own. Still, wishing to help those who were qualified to study but did not wish to do so, the Duke of Cambridge introduced a loophole. As in France, they were allowed to enter the staff by sitting for the final examination. As in France, too, and for much the same reasons, very few officers bothered to avail themselves of this opportunity.

The college's fortunes during the next three-quarters of a century were mixed. Its greatest success probably consisted in attracting and developing some exceptionally fine faculty; prominent among them were Colonel Edward Bruce Hamley, Lieutenant Colonel Charles Chesney, and Colonel G. F. R. Henderson, men whose military writings were the equal of anything produced during the second half of the nineteenth century.[37] This fact notwithstanding, the reforms were unable to retain their momentum. Though the requirement that the students pay for the course had been dropped, attendance at Camberley remained expensive. At a time when French and German officers drew increased pay for attending their respective staff colleges, English ones had to add perhaps a hundred pounds a year over and above their salaries to live in anything like the expected style. This may help explain why the number of those who applied for a place decreased from fifty-six in 1858 to twenty-three in 1867, a bad sign.

As is so often the case, the problems experienced by the college were as much the product of the surrounding environment as of its own structure. In an army that was, and still is, famous for the strength of its regimental system, attendance at Camberley had the effect of breaking off an officer's career and severing his connections. A general staff corps, capable of absorbing graduates and looking after their interests, did not exist until it was established in 1906–7 as part of the Haldane reforms that followed the Boer War. Consequently, staff officers were often left dangling in midair, deprived of the kind of patronage that, even after purchase was ended in 1874, remained vital for promotion.[38] The college therefore tended to attract middle-aged officers whose careers had become stuck (in 1885, the maximum age of admission had to be reduced to thirty-seven) rather than young hotshots on their way to the top. A brief check on those who attained senior rank *without* passing through the college confirms this interpretation. The list reads like a Who's Who of late-Victorian and -Edwardian soldiers; it includes Garnet Wolsley, Evelyn Wood, Frederick Roberts, Redvers Buller (who entered but never completed the course), Horatio Kitchener, and John French, field marshals all.

Academically, too, the program had its problems. The system of holding entry examinations notoriously led to cramming and superficial learning, though this was a problem from which not even the Kriegsakademie was exempt.[39] Another problem which Camberley shared with similar schools, particularly of a later date, was the tendency of the curriculum to expand until it lost its focus; it turned into a smorgasbord of different subjects including, in one form or another, military history, administration and law, strategy, artillery, fortifications, surveying, reconnaissance, mathematics, foreign languages, natural and experimental science, and riding.[40] The program's excessively theoretical character, with its emphasis on mathematics, reasserted itself. Military studies suffered comparative neglect, with the result that the first visit to an actual battlefield only took place in 1881.

Nor was the general intellectual atmosphere, in- and outside the army, favorable to serious study. In the age of the gentleman and of the self-proclaimed amateur, one did not advance one's career by reading (much less writing) books or by displaying excessive intellectual zeal.[41] This approach was fostered from the top when the Duke of Cambridge, who had done so much to establish the college, turned against it. Testifying before a parliamentary committee in 1870, he argued that the best staff officer was the regimental officer.[42]

Finally, British society at the time did not share either the French taste for *la science* or Prussian-German respect for *Bildung*. It looked on war as neither a science nor an art but as some exhilarating if slightly dangerous open-air game.[43] They were lucky in that, throughout the nineteenth century, their kind of war was waged almost exclusively against "natives" in faraway colonies. These opponents were regarded as something little better than the game British gentlemen also liked to hunt and to which they were also supposed to apply the rules of "fair play."

The view of war as "the great game" (this was actually the nickname given to the endless round of minor hostilities on India's Northwest Frontier) was popularized by writers such as Rudyard Kipling and the young Winston Churchill. It was also deliberately fostered in the public schools, where the vast majority of future officers were educated. Games such as cricket and football were surrounded by what can only be called a mythology whose express purpose was to explain their similarity to war and consequently to justify their prominent place in the curricula.[44] The team spirit that caused boys to "play up" in a game of rugby was celebrated in song as identical to the one that had won the empire in countless campaigns. Needless to say, a society that looked on war as an amusing extension of either hunting or football saw it as the opposite rather than as the continuation of study.

To quote Sir William Robertson, who was chief of the imperial general staff from 1915 to 1918 and who had himself served a tour of duty as commandant, the greatest achievement of the nineteenth-century staff college was probably that it provided "mutual agreement and excellent comradeship"[45]—and this was intended as a compliment! Change only started to get under way during the last years before World War I. What was seen as the army's failure in the Boer War resulted in widespread public criticism and provided an opportunity for reformers. They may have been helped by the fact that even the navy, always considered the senior service, was entering the twentieth century. It came to see the need for a general staff and for some kind of institution to train officers to serve on it.[46] In 1906, for the first time, two officers representing the world's proudest naval service were detailed to join the course, a clear sign that things were on the mend.

In 1906–7, as already noted, the army was finally provided with a proper general staff following the Haldane reforms. The effect was to provide the college with a new purpose in life, and the number of candidates for the entry-examination jumped in consequence. By 1914 both the chief of the imperial general staff and his deputy were graduates, a highly effective way of advertising Camberley's newfound importance. The last remaining administrative links with Sandhurst were severed and the post of Commandant was upgraded from colonel to general, thus formally putting Camberley on a par with the Kriegsakademie and the École Supérieure. In 1906 King Edward VII was persuaded to pay the college a visit, the first ever by a reigning monarch.

The curriculum was also reshaped to shed its eighteenth-century characteristics. It now laid less emphasis on mathematics and more on training in staff duties as well as war games complete with full dummy orders and schedules that had to be prepared with meticulous accuracy. Great emphasis was placed on developing the ability to summarize complicated situations in brief messages, though one general was to comment that the effect was not realistic since no attempt was made to make students perform under pressure of time.[47] All in all it should be said that the Haldane reforms, which in 1914 gave Britain the finest army ever to leave the island's shores, also helped ensure that the British high command in World War I was probably as good (or as bad, depending on one's point of view) as that of any other major belligerent. Unfortunately they came much too late to give the army a general staff corps commensurate with its size, which during the war grew almost twenty times over.

Compared with the reforms of the Haldane era, which violently pulled the British Army into the twentieth century and for a time gave the staff college an exceptional sense of purpose, nothing very important changed at Camberley between 1919 and 1939. The system of entrance examinations remained in force, theoretically requiring a year's preparation but often accomplished with the aid of crammers. The age of the students continued to drop until most were in their early thirties, a sign that the principle of using the college (or, at any rate, admission to it) as a vehicle for selection to medium rank had at long last become firmly established. The program continued to shed the remains of its theoretical bias, now widely recognized as antiquated. In keeping with the lessons of prolonged stationary trench warfare, it emphasized general staff work and military organization and administration, all carefully graded to progress from the divisional to the corps and army level. The second year included a foreign tour to some World War I battlefield and large combined exercises with the RAF and navy, a field in which the college was ahead of its time.

On the negative side, there was the fact that the faculty now often consisted of experienced veterans. Naturally they took the view that the only military history that mattered was recent history and that future conflicts would be decided by methods similar to those with which *they* had fought and won World War I, thus helping delay the switch toward mechanization and armored warfare.[48] More surprisingly, the image of war as a predominantly military affair remained unchanged. Though the program now included visits to large industrial firms, Britain had nothing like the U.S. Army's Industrial College and no systematic attempt was made to equip officers with an understanding of the nonmilitary aspects of war. Nor was Camberley's standing enhanced by the attempts, which got under way in the late twenties, to superimpose another tier of military education above it in the form of the Imperial Defense College.

Following the entry into the Staff College of many highly experienced officers, all tests and passing-out examinations were abolished. They were replaced by a final evaluation, couched by the Commandant in gentlemanly tones and seldom having anything negative to say concerning an officer. Camberley became what it still is, namely a highly cultivated place where the atmosphere is pleasant and the food very good. Here officers who are also "good sports" spend a non too strenuous two years playing games, thinking about war, all the while consuming large quantities of alcohol and socializing among themselves.

Russia–Soviet Union

What was ultimately to develop into the Soviet Union's principal staff academy was originally set up in 1832 as part of the military enlightenment, which had, at long last, reached Russia. Superficially the two-year course resembled that which was offered at similar institutions abroad: subjects taught included artillery, tactics, topography, geography, castramentation, logistics, strategy, military history, staff work, and foreign languages (German and French). Though all this sounded fine in theory, a look at the student body proves otherwise. In selecting their students, the Russians did not even pretend that they consisted of experienced officers designated for senior posts in the army. Instead the academy accepted junior officers over eighteen years of age who had to be of noble descent—which in practice meant that it became a playground for the spoiled sons of the aristocracy. Consequently it resembled the old Frederican académie des nobles than more than it did its descendant, the Berlin Kriegsakademie.

Surrounded by an institution that was not exactly noted for its intellectual prowess—the tsarist army—the General Staff Academy was not a success, all the more so because it came under French influence (its founder was the one-time French officer and military critic Antoinne Jomini) and concentrated on theoretical science. There were, to be sure, some attempts to correct its shortcomings. After Russia's disastrous military performance in the Crimean War, the place of general science in the curriculum was reduced in favor of what was then known as tactics and strategy and has since been renamed the operational art of war. In 1869, shaken by the Prussian success against Austria, the army was persuaded to set up separate advanced professional schools for engineers and artillerymen; this left the academy free (in theory at any rate) to focus on general staff work and on what was then understood as the higher conduct of war. It was in 1869, too, that the academy was renamed after the reigning tsar's father, the Nikolayevskaia Akademia, a sign of royal favor and of increasing importance.

From 1832 to 1918, the year it was dissolved, 4,532 students passed through the academy, a figure that includes those who went through an accelerated program during World War I.[49] Its contribution to the effectiveness of the tsarist army was limited; the task of the Russian school, like that of its counterparts in France and Britian, was merely to train deskbound officers for general staff service, and it did not succeed in establishing anything like a monopoly over promotion to high rank. Though the academy did publish some texts on strategy, it never set itself up as a recognized center of military-theoretical excellence. What it did do was create a body of officers who, though perhaps not brilliant, were at any rate *au courant* with existing military doctrine and willing to put that knowledge at the disposal of the young Red Army. Among them were Vatsetis, who made his mark primarily as a theoretician; Bonch-Bruevich, chief of staff to the revolutionary high command; Kamenev, who rose to become a front commander; and Shaposhnikov, who was later to serve as Stalin's chief of the general staff.

The Bolshevik leaders who created the modern Soviet Union were, whatever their shortcomings, perhaps the most intellectual group of men that ever held power in a large country. They were also acutely conscious of the educational backwardness of their fellow citizens and determined to correct it. Therefore, and even though the years of political disintegration, civil war, and famine that followed immediately after 1917 were scarcely conducive to study, right from the beginning they insisted that the existing system of advanced military education be not only retained but expanded. At a meeting held on March 10, 1918, Lenin personally drove through the decision to provide the new Red Army with a staff academy. A supervising committee was set up, and in the midst of chaos it was able to do sufficient useful work for the reconstituted academy to officially open its doors in November of the same year.

The period of study at the new academy was initially set at one year, but this proved impossible to achieve owing to the turbulence of the times. The goal of creating a school that would be on a par with the best anywhere was not lost, however, and as soon as the civil war was over a three year curriculum was instituted. The faculty consisted of former senior tsarist officers, some of them of general rank. Ideological questions were taught by political activists and visiting speakers from the party elite. Student-selection procedures stressed party loyalty first and military proficiency second. Among the initial batch of 183 students, 90 percent were personnel who already held a commission and who had gained practical experience during the past five years. Their military qualifications were thus as good as those of any similar group in the West; however, the entry of new social groups into the officer corps meant they could not compare in general development. Only 5 percent had spent any time at an institution of higher learning, whereas the education of fully 25 percent consisted only of elementary school or less. The academy tried to correct this deficiency, which in some cases extended to the basics, by providing remedial courses for those who needed them. The task of bringing the Soviet officer corps up to par took time, and only began to bear real fruit during the nineteen thirties—just as Stalin's purges got underway and uprooted much of what was best in it.

As in every staff college around the world, a very heavy workload was put on the students, who were expected to spend as many as eight to nine hours daily receiving formal instruction in class, though this was later reduced to seven to eight. Most of that time was devoted to political-social-economic studies—which were not yet known as Marxism-Leninism—military history, and military art in that order. Military art in turn consisted of strategy, the tactics of the various arms (artillery, cavalry, engineering), organization, logistics, transport, and administration. Except for the absence of air warfare, which had to be added later on, this was a modern curriculum by the standards of the time.

July 1921 saw the appointment as superintendent of M. K. Tukhachevski, the man who had just come within a hair of occupying Warsaw but who was—through whose fault is still moot—finally defeated and thrown back into the Ukraine. His coming opened a process by which the Academy's status was raised until it reached the elevated status which it occupies today. Renamed "The Military Academy of the Workers' and Peasants' Red Army," it became a combined warfare school that prepared officers for commanding formations from brigade level up. When Tukhachevski left to become deputy minister of defense in 1924 he was replaced by M. Frunze, the USSR's leading military intellectual. He introduced a shift from theoretical to practical study, as well as a heavy emphasis on wargames and outdoor work. Among those who taught at the academy during this period were a future deputy chief of staff, V. K. Trianfilov, and a future inspector general of the armored forces, K. B. Kalinovski.

The thirties saw a massive growth of the Soviet armed forces, a shift from territorial to cadre divisions, and a growing realization that another war could not be long delayed. These three developments together threatened to swamp the military academy and led to another reorganization. Specialized arms training—artillery, engineering, signals, the lot—was taken away from the military academy. Instead it was entrusted to a series of thirteen newly established arms schools *cum* staff academies that have no exact equivalent in Western armies. The military academy itself was

One of the first to conclude that commissioned officers might benefit from further training and education was Prussia's Frederick the Great. Between one war and the next he attempted to provide his officers with food for thought, pouring forth a stream of instructions, testaments, and *Kriegsgesaenge* (warlike songs). (M. Lehman, *Scharnhorst* [1886], Courtesy Library of Congress)

The real founder of modern advanced military training was General Gerhard von Scharnhorst (1755–1813). His *Kriegakademie* was the best of its kind by far and served as a model for all others. (H. von Petersdorff, *Friedrich der Grosse* [1911], Courtesy Library of Congress)

The mainstay of the British "Senior Department" during the first half of the nineteenth century was John Narrien. A mathematician by profession, he went blind in old age. This did not prevent him from teaching topography, artillery, and fortification as well as conic sections. (Courtesy the Camberley Staff College)

During the Crimean War the British Army suffered from gross deficiencies in staff work and organization which were brought to the public attention by means of telegraph and press. There followed a reorganization of the Camberley Staff College, which in turn led to the above cartoon in the satiric magazine *Punch* showing a field trained officer (on horseback) confronted by the new academic requirements for promotion, personified by the gentleman/scholar (on foot). (Courtesy *Punch* Publications Ltd.)

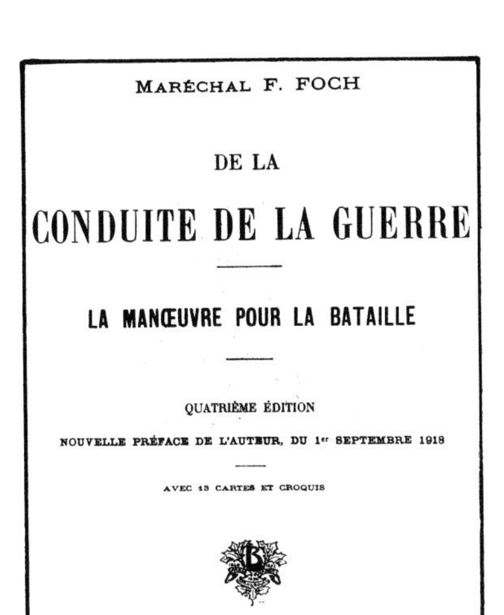

MARÉCHAL F. FOCH

DE LA

CONDUITE DE LA GUERRE

LA MANŒUVRE POUR LA BATAILLE

QUATRIÈME ÉDITION

NOUVELLE PRÉFACE DE L'AUTEUR, DU 1er SEPTEMBRE 1918

AVEC 13 CARTES ET CROQUIS

BERGER-LEVRAULT, LIBRAIRES-ÉDITEURS

PARIS — 5-7, RUE DES BEAUX-ARTS | NANCY — RUE DES GLACIS, 18

1918

Foch's *La conduite de la guerre* was originally written as a textbook for his students at the *école supérieure de guerre*. It was not before the fourth edition of 1918, whose title page is reproduced here, that the author so much as deigned to mention the role which industry plays in war. (F. Foch, *La conduite de la guerre* [4th ed., 1918], title page)

Probably the greatest strength of the old German *Kriegakademie* was that its instructors, such as General Erich Ludendorff, were promotable officers on the way to the top who could serve the students as role models. (Courtesy the Bibliothek fuer Zeitgeschichte, Stuttgart)

During the Second Reich, General Staff officers were known as *die halbgotter* (the demigods). In 1913 *Simplicissimus*, a satiric magazine akin to *Punch*, published this cartoon of their life. The caption reads "In the Staff officers' mess: 'not bad, comrade, considering how far our situation is supposed to have deteriorated.' " (Courtesy *Simplicissimus*, 1913)

Captain Alfred Thayer Mahan was probably the greatest military-theoretical mind ever to teach at a modern military institute of higher learning. Even he failed to make the Naval War College into a major planning center or to turn it into a vehicle for selecting and promoting personnel. (Courtesy Naval War College, Newport, R.I.)

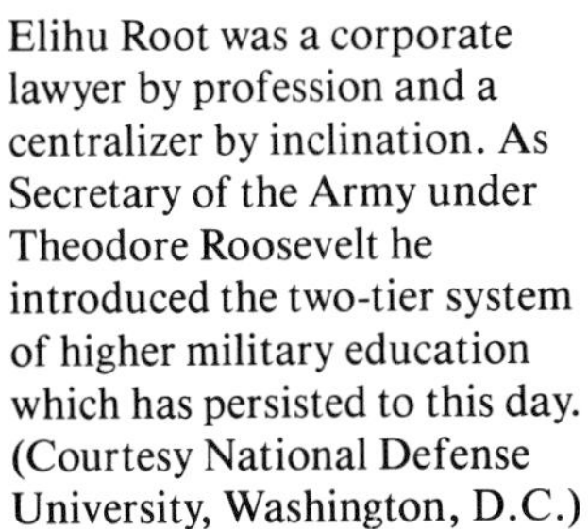

Elihu Root was a corporate lawyer by profession and a centralizer by inclination. As Secretary of the Army under Theodore Roosevelt he introduced the two-tier system of higher military education which has persisted to this day. (Courtesy National Defense University, Washington, D.C.)

The preferred method of instruction at the Naval War College during the thirties was the war game. Admiral Nimitz later claimed that, with the exception of the Kamikaze tactics, "absolutely everything" the Japanese did in World War II had been foreseen in those games. (Courtesy Naval War College, Newport, R.I.)

The U.S. Army General Staff and Command College at Fort Leavenworth has always been a training school first and an intellectual center second. During the thirties even the Army War College faculty complained of the "Leavenworth mind" and it's preference for rote learning. (Courtesy National Archives, Washington, D.C.)

The National Defense University at Fort McNair, Washington, D.C., is theoretically the highest institute of military learning in the Western world. (Courtesy National Defense University, Washington, D.C.)

The intellectual and theoretical inclinations of the leaders of the Russian Revolution resulted in more serious attention being given to military studies. Here, a group of students in the Soviet Union pursue such studies at the Frunze Academy, around 1960. (Defense Ministry, USSR, ed., *Fifty Years Soviet Army*, Moscow, 1967)

Anyone who has read Soviet military literature will be familiar with Communist platitudes in the form of obligatory quotations. However, Marxism-Leninism can also serve as a framework for military analysis. The photo shows a political-education class at a Soviet staff college. (Defense Ministry, USSR, ed., *Fifty Years Soviet Army*, Moscow, 1967)

The Afghanistani mujahedeen, like the Viet Cong, are not known to attend any staff or war college, let alone to earn university degrees. Though neither were "professional" soldiers, they were effective enough in fighting the greatest military powers in the world. Here, a shell-laden guerrilla balances atop a treetrunk serving as a bridge. (Reuter/picture N. 1SL02 Afghan by Eth/Mwzammi Fasha, June 1, 1988)

renamed the Frunze Academy. Though formally its standing is similar to that of the remaining arms schools, in practice it became the USSR's most prestigious combined warfare school, with attendance a virtual must on the way to senior rank. The task of offering instruction in the higher conduct of war was, however, taken away. It became the province of a new institution, the General Staff Academy, which from 1936 on was to occupy top rank in the Soviet military education pyramid.

Most of the instructors at the new academy were brigade and division commanders, which meant that they outranked their opposite numbers anywhere else. The students were colonels and lieutenant colonels who had already graduated from Frunze (rarely, one of the other arms schools) and who were clearly destined for senior posts, either as divisional commanders or on the staffs of corps and armies. The two-year curriculum included the usual political indoctrination; however, it also put a heavy emphasis on strategy, interservice organization and cooperation, military history, and the nonmilitary aspects of war, including industry, transport and mobilization. The Academy was unusual in that the students spent only three days of each week in class, the remainder being devoted to independent work. The last three months of the program were set aside for the writing of a thesis on which graduation depended, this being another feature that had no parallel elsewhere. When the Germans attacked in 1941, six hundred men had passed through the academy's doors. Its importance in the Soviet order of things is perhaps best brought out by the fact that, unlike its equivalents in other countries, it was never completely closed down even during the worst years of the war. Nor had hostilities even come to an end before the full two-year curriculum was restored.

There has been, in the West since 1945, a tendency to belittle the quality of the Soviet military, including the quality of military education.[50] Much of this can be traced back to the German generals, whose memoirs served Westerners as a key vehicle for understanding the War on the Eastern Front. Eager to win recognition, both for themselves and for their country as part of NATO, a generation of ex-Wehrmacht commanders developed an institutional interest in attributing Soviet achievements—and their own defeat—to quantity rather than quality.

Perhaps even more important, Soviet quantitative superiority represents one "permanently operating factor" that has lasted to this day, or so the common wisdom goes.[51] Western military establishments have, of course, made use of this alleged superiority in order to demand additional funding for themselves. At the same time, they helped develop the view that Soviet quality just *had* to be inferior; or else NATO's entire politicostrategic rationale, and with it any chances that it stands of surviving in war, would crumble into the dust.

Against this disparaging view of Soviet intellectual capacities before and after World War II, the following facts stand out: Lenin, Trotsky, and even Stalin—all of them following the tradition first established by Engels and endorsed by Marx—took a serious interest in military affairs, studied them, and wrote on them, which is more than can be said about most civilian leaders in most other countries. Well aware of their country's backwardness in many respects, including military education, they did what they could to correct it. The training system that resulted was built on the foundations laid by the tsar's regime. Structurally it was somewhat different from that of other countries in that arms schools and staff academies were united (after 1930) in single institutions and conducted by each arm separately. The course at the most important academy, the one named for Frunze, lasted three years, which was longer than in any other country except Germany. Moreover, after the General Staff Academy was opened in 1936, senior Soviet officers actually spent more years at school than did their opposite numbers anywhere else.

Admittedly, it is not clear even today how schooling and practical experience should be balanced against each other in the training of senior commanders. Then, too, there is no way to compare the quality of instruction at Soviet institutes of higher military learning with that offered elsewhere. Still, there is no mistaking the seriousness with which the Soviet leadership approached the task or the massive scale of the effort that went into it. Finally, results count. The Soviet high command may not have performed too well during the Russo-Finnish War or during the first year of World War II, nor did the quality of junior- and medium-level Soviet leadership ever stand comparison with that of their German opponents. In spite of what many see as Stalin's murderous interference, however, later during the war the performance of the Soviet high command was second to none, including even the vaunted Germans.[52] Though it would be too much to give the General

Staff Academy sole credit for this achievement, on the other hand it certainly does deserve some of that credit.

United States

In the United States as in so many other countries, serious concern with advanced officer training and education first arose out of the spectacular Prussian victories in 1866 and 1870–71. In 1875, persuaded that the U.S. Army had fallen behind, the secretary of war sent the then–Brigadier General Emory Upton, a much-decorated veteran of the Civil War, on a fact-finding journey around the world. Something of an intellectual himself—he had been an instructor at West Point—Upton had specific instructions to look into the question of advanced military education and particularly in the German Kriegsakademie. His return to the United States was followed by the publication of a book, *The Armies of Asia and Europe.*[53] There he complained about the intellectual backwardness of the American military and noted that U.S. officers, artillerymen only excepted, did not receive any form of advanced training. He therefore recommended the founding of infantry and cavalry schools, as well as a staff college on the German model.

During the years after the Civil War, the U.S. Army was a minuscule organization not to be compared with the forces of any other major power. To make matters worse, it was scattered in penny packets all over a gigantic continent. Its main function consisted of fighting the last remaining Indian tribes, a task that had everything to do with minor tactics and little with study of any kind. There was a commander in chief (not a chief of staff), but real power was concentrated in the hands of the various arms, each headed by its own bureau. Though the bureaus were quite willing to provide officers with education, they were even more determined to preserve their own autonomy. This led to the founding of a separate school by each arm. A School for Application of Light Artillery (note the French influence here) was opened at Fort Riley in 1869. In 1881 a School for the Application of Cavalry and Infantry was opened at Fort Leavenworth, Kansas, Colonel Elwell S. Otis commanding. However, the bureaus drew the line when it came to a question of establishing a unified school for the entire service.

The early years of Leavenworth were, intellectually speaking, anything but edifying. The post's remote location prevented much contact with the nation's principal intellectual centers, all of which were located in the Northeast. An old warhorse himself, General Sherman in his capacity as commander in chief had little use for theoretical study. It was his view that "the school should form a model like Gibraltar [*sic*], with duty done as though in actual war, and instruction by books be made secondary to drill, guard duty, and the usual forms of a well regulated garrison."[54] Accordingly he provided Leavenworth with actual troops, to wit four companies of infantry, four troops of cavalry, and one battery of light artillery, to be commanded by instructors and students in addition to their educational duties. What time this left for "academic" work was devoted largely to remedial classes in writing, grammar, and mathematics. The military part of the curriculum was limited to minor tactics, initially taught with the aid of foreign textbooks since none were as yet available in the United States. If only because the army still had no general staff, staff work as a separate specialty was neither recognized nor taught.

Since educational reform in the army was stalled, the initiative passed to the navy. It was prodded by another intellectual officer, Captain (later Rear Admiral) Stephen B. Luce, who was impressed by the fact that the navy did not even have the kind of applicatory schools recently established by the army. For Luce, success in war demanded more than "the application of electricity to torpedoes, chemistry to explosives, or metallurgy to ordnance";[55] officers were to receive instruction in "the higher branches of the naval profession," including "the science of war, naval tactics, military and naval history, international law, military and naval law, modern languages, and such elective branches as might be found desirable." A seven-year campaign involving both letters to the secretary of the navy and articles published in the press initially made little headway. Finally, in February 1885, the U.S. Senate got involved and ordered the navy to go ahead. Seven months later the Naval War College was opened, with Luce himself acting as both president and instructor—a first ever for any navy.[56]

The victory thus won was, nevertheless, a very partial one. Late-nineteenth-century naval officers, American as well as foreign, were nothing if not crusty seadogs. They were accustomed to commanding their ships as if they were God running the universe—these were the days before the introduction of radio—and found it hard to accept that anything useful could be accomplished by sitting behind a desk and studying. This negative attitude was reflected by the fact that the college was located in a building at Newport, Rhode Island, which had originally served as a poorhouse *cum* deaf-and-dumb asylum. No money was provided for books, furniture, heating, or even light; Mahan later remembered carrying the one available lamp from one classroom to the next. Strangely enough, the only full-time faculty member was an army lieutenant charged with teaching military history—not exactly a normal situation for an institution whose declared purpose was to instruct officers in the higher conduct of naval warfare.

Luce had wanted to copy the European system of governing entry to the college by an examination, but this too he failed to achieve. Instead, each year a handful of students—lieutenants and lieutenant commanders—were ordered to Newport by the navy in the same way as they received any other assignment. Courses were at first limited to six weeks during the summer months; twenty-five years would have to pass before the duration of the courses was extended to cover as much as an entire year. Overall, the college was tolerated rather than accepted. The fact that it survived the attempts to close it down or to merge it with the nearby torpedo school was largely due to the work of one of its lecturers, Captain Alfred Thayer Mahan, whose writings were to turn it into a center of national, and even international, attention during the nineties. Still, the navy was not impressed. Mahan's books were used as propaganda for building an oceangoing fleet of battleships, but the college never became the professional planning center its founder had intended. Nor did it play an important role in the selection and promotion of naval officers.[57]

When the Spanish-American War broke out in 1898, the U.S. armed services were swamped by operations on a scale for which they had never been designed. Mobilization, logistics, and transport to the theaters of war were characterized by a vast mess. The bad publicity that resulted was all the greater because most of those affected were national guardsmen and civilian volunteers—in other words not just a small group of socially isolated regular soldiers. The way toward reform was opened when Theodore Roosevelt took over the presidency in 1901. The man he put in charge was the secretary of war, Elihu Root, whom Roosevelt had inherited from his murdered predecessor McKinley. A corporate lawyer by profession, Root was a centralizer by inclination. His principal concern was to end the dominance of the bureaus and create the machinery by which a much-expanded army, now engaged in active operations overseas, could be managed.

As part of this drive toward centralization at the top, Root initiated an investigation of the old School of Application for Infantry and Cavalry at Fort Leavenworth. The result was much as could be expected: What the United States had was more like a remedial school for semiliterate officers than a proper staff college, nor did it even offer joint training to army officers of the various arms. With the dominance of the bureaus finally broken, the last-named defect was comparatively easy to correct. By degrees between 1902 and 1907, attendance at Leavenworth was expanded. Officers from the artillery, signal corps, the engineers, and the medical corps were brought in. The school itself was reorganized until it offered a one-year course with a strong emphasis on combined arms training and staff work—that is, subjects that were common to the army as a whole rather than limited to any single arm.

Apart from the period 1927–36, when the duration of the course was extended to two years, little changed at Leavenworth between the time Root left office and the outbreak of World War II.[58] The students were mostly captains in their early thirties who had already passed through the school of the arm to which they belonged. They were selected on the basis of efficiency reports and the recommendation of their superiors, rather than by competitive examination as was the system everywhere else. To have one's name entered on the published list of entrants was considered a very important step on the way to promotion. However, attendance at the college was never made an absolute prerequisite for advancement.

The college's declared task was to prepare staff officers capable of effective service at divisional, corps, and army headquarters. Its curriculum was divided into two major parts, one dealing with the preparation for war and the other with its conduct. The first covered the functions and operations of the four (initially, five) staff departments (Gs), which were copied from the French and assumed their present form after the end of World War I. The second consisted of playing war games—particularly the "free" version taken over from the Germans—and preparing plans for hypothetical wars. The faculty, like the curriculum, was almost entirely military. It consisted mainly of majors who impressed their superiors as more didactically inclined than the rest. Normally they did not spend a long period in the assignment, hoping to proceed to higher things. With such a body of instructors the general staff and command school became even more of an intellectual backwater than was dictated by its remote geographical location. During the thirties even the faculty at the Army War College, themselves no great luminaries, were to complain of the "Leavenworth mind" and its tendency towards rote learning.[59]

All in all, Fort Leavenworth graduates are perhaps best described as pale copies of the lieutenant colonels produced by the Berlin Kriegsakademie. We may take it that they qualified to serve as medium-level military technicians, staff officers, and administrators. There was no question, however, of the one-year course producing operational experts on the German model. The curriculum did not introduce officers to the political, social, and economic problems involved in the higher conduct of war, nor did it do much to teach Americans about the habits and thought processes of the foreign nationals they might one day be required to fight.[60]

The establishment of the command and general staff school was only the first reform Root introduced into the officer-training and education system. To paraphrase the title of a book that is said to have influenced his thought,[61] his purpose was to provide the army with a brain on the Prussian model. The outcome was the founding of the Army War College as the highest institute where officers would be trained in the conduct of war. Its location was in Washington, D.C., on a peninsula between the Potomac and Anacostia rivers that had previously served as an arsenal and where Lincoln's murderers had been hanged. There $700,000 were spent to build a modern campus, later renamed Fort McNair.

The college's first commander was Brigadier General Tasker Bliss, a well-educated officer (he had done translations from the classics) whose qualifications included service as an instructor at West Point and at the Naval War College. As he saw it, the task of the new college was not to duplicate the work being done at Fort Leavenworth; instead, it was to function as a kind of preparatory general staff. Here students, working under supervision, would be entrusted with tasks normally entrusted to the newly created War Department general staff, particularly its planning division.

Root himself assigned the college its first task—drawing up a detailed inventory of America's armed forces, their location, equipment, and preparedness for war. Subsequent assignments included the preparation of plans for the withdrawal of US forces from the Philippines, the defense of the border with Canada in case of a war against Great Britain, and for war against Mexico. However, the arrangement by which the college's commander wore a double hat as chief of the planning division proved as unworkable in the United States as it did in France. Though the formal connection between the college and the War Department was not severed until 1919, already in 1905 the curriculum was revised to devote more time to lectures and exercises and less to actual planning.[62]

The two decades before 1919 also saw the establishment of the Army War College's other salient features, many of which have continued to characterize it to the present day. The duration of the program, originally fixed at four months, was extended until it reached one year. Like Leavenworth, but perhaps with greater justification owing to the students' greater seniority, the college did not make admission dependent on an examination. Unlike Leavenworth, it did not even make a pretense of rating its students. In 1919–21 the commandant started classifying them as either "fit" or "not fit" for service on the War Department general staff. However, this presumptuous attempt was rejected almost immediately by the chief of staff, who saw his own prerogatives undermined.

Nor is it very clear who should have graded whom, given that students and faculty tended to become more and more like each other. Originally it was a case of majors teaching captains, but as the maximum age of admission was reduced from fifty-two to forty-four, both groups gradually came to consist of colonels and lieutenant colonels, with the occasional major thrown in. The fact that there was no discernible difference between students and faculty may help account for a growing tendency to rely on guest speakers. This was all the more the case because the college's location in the heart of Washington, D.C., made such a policy both attractive and easy to implement.

When Root founded the college he had intended it to be a high-visibility, high-prestige institution that would attract the Army's best. This policy was implemented when General Bliss, the college's founder and first president, went on to become chief of staff; none of his successors was able to repeat anything like that feat, however, and the post gradually diminished in importance until it became a sinecure on the way to retirement. Nor did even the most respected commandants ever succeed in gaining any real control over the selection of faculty and students. About the only prerogative they did have was to tinker with the college's internal structure and with its curriculum, which more than any other factor may account for the frequent changes in the latter.

The interwar years saw any number of reorganizations and changing emphases, now on "general staff work," now on "the art of command," now on general education, or military history, or war games. Meanwhile the number of lectures, which Root had wanted to keep to a minimum, grew and grew until students were subjected to no fewer than 130 a year. Such a number left little time for independent work. It may have had something to do with the failure, in 1936–37, of the first of several attempts to give the college university status and thus enable its graduates to take an advanced degree.

On the plus side, the U.S. Army was among the first to grasp the implications of World War I for the conduct of future armed conflict. In 1919, at a time when the authorities in most countries were only beginning to wake up to the problem, the American Secretary of War N. D. Barker wrote that modern officers required "a full comprehension of all agencies, governmental as well as industrial, involved in a nation at war." This view was adopted by the first commandant of the postwar college, General James McAndrew. Though implementation was often hampered by the lack of a qualified faculty, the curriculum showed growing concern with foreign policy, national objectives in war, the coordinated use of military and economic power, and the role of public opinion.[63]

Proceeding further along the same road, the US Army also realized that some future senior officers would require training in such fields as economics, business administration, and industrial technology. Supported by Bernard Baruch, the years from 1920 saw various attempts to meet this requirement. In 1924 they led to the founding of the Army Industrial College, whose task was "to educate officers in the useful knowledge pertaining to the supervision of procurement of all military supplies in time of war and to the assurance of adequate provision for the mobilization of material and industrial organization essential to war-time needs."[64] The first class consisted of just nine students, all of them reservists. It was a sign of the college's growing importance, however, that regular officers were sent to attend it from 1930 on. By 1939 the number of students had grown to sixty-two, including some national guardsmen as well as representatives from several government departments.

To judge by the results achieved in World War II, the Army Industrial College was a smashing success. This success cannot have been due to the profundity of the course, which, after all, lasted just one year and suffered from the usual shortfall of qualified faculty. Nor did it have anything to do with the method for grading the students, for no such method existed. Rather, three factors may have been involved. First, during all of the twenties and much of the thirties, "business" was considered a highly prestigious occupation, well worth the practical man's while; at the same time it was sufficiently new to virtually all officers for them to approach it with an open mind. Second, the college was able to form useful contacts with such government departments as State and the Treasury, which were later to play a vital role in the higher conduct of the war. Finally, the college benefited from the fact that the subject studied was not military in the narrowest sense. This caused the air corps and even the navy to relax their mutual antagonism, thus facilitating intellectual cross-fertilization.

While the army engaged in a slow process of partial self-reform from the eighteen seventies on, the navy, as usual, went its own way. In 1919 a special board was appointed to look into the question of advanced training and education for officers; one of its members was Ernest J. King who, at that time, held the rank of captain.[65] They submitted a carefully thought-out proposal for a three-tiered, graduated program of instruction designed to carry the Naval officer from subaltern to flag officer. The program was accepted "in principle." However, the navy argued that it did not have enough officers to permit the luxury of extensive periods of study.

During the interwar years, the normal method of training and educating naval officers was as follows. Having gone through their first sea duty, they were called to attend a so-called postgraduate course at Annapolis, lasting one year. Upon graduation, some of the officers proceeded directly to a second year of study at a civilian school of engineering whereas others attended the navy aviation, submarine, or torpedo school. Fifteen years after being commissioned officers, who by this time would be either commanders or lieutenant commanders, might be selected to attend a one-year junior course at the war college at Newport. After another five years' service, those who had reached the rank of captain might be called to the senior course—though in practice the contents of the two courses were alike, so that very few officers ever went through both. At a time when the army made at least some attempt to use Leavenworth as a screening device, the navy insisted that shipboard duty was almost the sole highway to promotion. Consequently, those who attended the naval schools and colleges saw them mainly as places to get their tickets punched.

At the Naval War College, as indeed at all lower levels of naval education, the subjects studied were strictly naval and practical. This applied even to the senior course attended by captains in their mid-forties; by far the most important subject of instruction was "grand tactics," by which people meant roughly what is nowadays known as "the operational art of war." Instructors and students conducted make-believe maneuvers, grappled with "operational problems," and made plans for hypothetical wars, including the one that was ultimately fought against Japan. Much the most important method of teaching was the wargame. This involved officers moving model-ships on the liloneum-covered floor of a large room, and is said to have helped design tactics and enforce strategic realism.[66] By contrast, studying the past consisted mainly of endless very detailed rehashings of the Battle of Jutland in 1916. The U.S. Navy was at one with its British counterpart in regarding this battle as the greatest event in the history of the planet and as a model for others to come.

To sum up, the American system of training and educating senior officers was much more decentralized than the German one on which it was partly modeled and that led the world. There was in the United States no single institution of higher military learning that could compete, in the quality of study as well as in social prestige, with a first-class civilian university. Moreover, American culture in general, and the services in particular, did not share either the German ideal of *Bildung* or the French one of abstract mathematical science (let alone the British one of games). Consequently, advanced military training in the United States put considerably less emphasis on theory—including military theory—than in those countries. The result was senior commanders who were doers rather than thinkers. Under the American system, neither a

Moltke nor a Schlieffen would have stood much chance of rising to chief of staff. To be fair, it should be added that officers of the wine-drinking, easy-going, "Papa Joffre" type would also have failed to reach high rank. Nor did the American system ever produce anything like Foch, with his single-minded insistence on *élan.*

The available evidence does not permit an evaluation of the quality of instruction offered at various American and foreign military schools and colleges. However, from the organizational point of view three facts stand out. First, no U.S. institute of advanced military learning relied on competitive examinations to govern entry, this being the result partly of a democratic tradition and partly of bureaucratic squabbles. Second, no such institute offered a comprehensive, systematic, integrated three-year course on military history, art, and science such as formed the core of the German Kriegsakademie and the true reason for its excellence. Third, and even though there were some slight differences between the Army and the Navy in this regard, no U.S. institute ever approached the importance of the Kriegsakademie as an instrument for screening and promoting officers on their way to top commands.

Taken in combination, these considerations help explain why no U.S. institute gained anything like the social prestige associated with its German equivalent—and to the extent that prestige helps attract quality, it is a factor that matters. They may also serve as one explanation, though admittedly a very partial one, as to why most U.S. Army ground commanders in World War II were not up to German standards at any level below army group.[67]

On the other hand, the United States enjoyed one great advantage their opposite numbers in other countries did not share. Possibly because the American system of top-level political management gave civilians considerable influence on the day-to-day affairs of the services, they were among the first to realize the importance of nonmilitary problems in the conduct of modern war. Increasingly as the thirties drew to an end, the Army War College and the Army Industrial College paid attention to questions of economic and industrial mobilization, acquiring some expertise in these fields and laying the foundations for cooperation with civilian agencies. In this respect they were matched only by the Soviet General Staff Academy—and even the Soviets did not have an equivalent of the Army Industrial College.

Furthermore, the Army Industrial College served a vehicle by which the forces established contacts with other government agencies that would be responsible for running another global war. These factors may help explain why America's top team of senior commanders in World War II—the George Marshalls, the Dwight Eisenhowers, the Douglas MacArthurs, the Ernest Kings, the Chester Nimitzes—were unmatched by anything on the Axis side. Going one step further, they may serve as one explanation—though again it is only a partial one—as to why much of the United States' grand strategy in World War II was superbly conceived and well executed. And why, as a result, the United States was able to become the world's leading superpower at the cost of only 300,000 dead.

Problems

Though the system by which medium and senior officers underwent some kind of formal schooling in order to prepare for their jobs was more than a century old in 1945, the idea that they should attend a university and earn a degree had not yet taken hold. Such famous commanders as Montgomery, Zhukov, or Koniev—not to mention Ludendorff, Foch, Schlieffen, or Moltke—did not hold degrees in military science or anything else. Academic degrees in general, and advanced degrees in particular, continued to be reserved principally to a handful of specialists such as engineers, lawyers, doctors, and priests.[1]

Even as late as World War II, U.S. armed forces were exceptional in that they contained a sizable sprinkling of officers who had graduated from college. This was due to two reasons. First, the American services had long offered the equivalent of a free college education—and the degrees that go with it—as a way to attract suitable personnel in a country that had neither a strong military tradition nor universal conscription. Second, once war broke out civilians were drafted on a massive scale. The principal instrument by which their military potential was determined was the written Army General Classification Test. Like all such tests, it discriminated in favor of those with a higher education.[2]

During the quarter century after 1945, for reasons that had very little to do with the military, higher education in all the advanced societies exploded. In a world that had abandoned God in favor of science, a degree came to be widely regarded as the next best highway to heaven. As a result, the number of Americans per 100,000 going to college or university more than doubled between 1950 and 1966, with most other countries doing as well or better.[3] Graduate enrollment rose from 105,000 in 1940 to 690,000 in 1967, and an average of 1,400,000 for each of the years 1980–84. The number of Ph.D.'s awarded in the United States grew from 3,300 in 1940 to 14,500 in 1964 to more than 30,000 in 1970, and stood at an average of 33,000 in each of the years 1980–84.

Other countries experienced similar developments. Wherever one looked, the number of new universities, the size of their faculties, the amount of research that they spewed out, and the share of GNP devoted to their support grew and grew. All expanded much faster than either the general population or national wealth, and indeed the rate at which they did so was often used as one index of modernization. Such was the rate of increase in higher education that, at one point during the late sixties, it looked as if everybody would end up by becoming his or her own professor.[4]

As the number of graduates grew, their position in modern society changed. To paraphrase Bismarck's quip about humanity beginning with the rank of lieutenant, anybody who was somebody now had to have a degree rather than a high school diploma as in the past. Conversely, anybody who did not have a university education was threatened by a drop in status, income, or both. One of the principal groups affected was officers—people who traditionally chose their career either because they were too proud to study for a degree or else (as cynics would say) because they were too stupid to do so. As the civilian world went to college and turned to "professionalism" as its ideal, a situation resulted in which officers became perhaps the only responsible segment of society that often had *not* received the blessings of a higher education. Increasingly, they faced difficulties when dealing with their opposite numbers in government and business—people who held not just degrees but advanced degrees.

Another factor that helped depress the officer's status—in the most advanced countries in particular—was the advent of nuclear weapons. Initially, the spectacle of A-bombs exploding over Japan and causing that country to end the war within as many days as the years it had previously lasted seemed to enhance the power of the military: It put them in a position in which they could literally rule over the world's continued existence. However, the final outcome was just the reverse. In every country that had the bomb, politicians regarded as intolerable a situation in which control over such weapons should rest in any hands but theirs. Accordingly, they were put under lock and key, or else carried around like so many barrels of wood that could be activated only by top-secret codes controlled from the National Command Centers of the countries involved. In East and West alike, the number of people actually in a position to activate nuclear weapons could be counted on the fingers of one hand. In East and West alike, the vast majority of soldiers of all ranks were turned into slightly pathetic figures free to play with conventional weapons, conventional being synonymous with second class.

The ever-present fear of nuclear escalation in turn meant that even conventional forces could not be employed in major conflicts among major powers, and indeed since 1945 no country that has the bomb has been engaged in such a conflict. Instead, their principal use—apart from saving lives and carrying out public works—was in so-called "low-intensity" and "counterinsurgency" conflicts against third-rate enemies. To cap the irony, the enemies in question often belonged to supposedly inferior races known by a variety of pejorative nicknames. Fighting such opponents, the military in the most advanced countries gained few kudos even if they won—which, very often, they did not.

With conventional war relegated to second-class status, the military in all the most advanced countries stood in danger of being turned into a mere constabulary. War as traditionally understood could no longer fill their lives, for fighting meant that deterrence had failed and that Armageddon might be immediately at hand. If the military profession was to maintain its place in society, its significance had to be altered. During the fifties *defense* and *security* gradually supplanted *war,* thus gaining a double advantage. First, unlike war, defense and security were continuous and could be presented as of overriding importance even in peacetime. Second, they included not only strategy (how to deploy one's forces), operations (how to maneuver them in the theater of war), and tactics (how to make them fight when in contact with the enemy) but almost every conceivable aspect of human existence. Soon hardly any field, be it national economics or the structure of whales' brains, was too esoteric for the military research and development establishment to meddle with. Conversely, any subject that did *not* interest the brass was, for that very reason, probably unimportant.[5]

Paradoxically, the downgrading of conventional war coincided with yet another trend. Though the end of World War II caused all the principal countries to demobilize, the belief that the next war would be as total as the last persisted for at least another decade.[6] Consequently, armed forces all over the world retained disproportionally large numbers of medium officers in service; the rationale being that, should another conflict break out, these officers would act as the backbone of an accelerated mobilization process. As the prospect of another total war receded over time, the group of phenomena collectively known as Parkinson's law took over. The proportion of "educable" middle-ranking officers (captain to colonel and their naval equivalents) rose and rose. To focus on the American example, it mushroomed from 35 percent at the end of World War II to 42 percent in 1946, 54 percent in 1949, 63 percent in 1961, 69 percent in 1973 (the year in which the Nixon administration put the forces on a volunteer basis), 65 percent in 1983, and a walloping 72 percent in 1988.[7] This created problems as to what was to be done with them.

Finally, the postwar world saw the adoption by many military establishments of a much sharpened-up or -out promotion system. It became mandatory for officers to reach a certain rank at a certain age or else face separation from the forces. This policy may have been justified from a military point of view, yet it did create social problems. Even as the life expectancy of the population rose, the culture became increasingly obsessed with youth. Taken together, these developments meant that retired officers could no longer be content with their predecessors' role as elderly gentlemen pottering around. Those of them whose only skills were military could expect trouble finding suitable jobs. This was all the more so because civilian higher education was exploding and flooding the market with degree holders.

Although things were by no means as clear in 1945 or even in 1975, as they have since become, providing officers with a college education turned out to be like the proverbial tailor who killed seven flies with a swat of his scarf; it solved all problems at a single stroke. Seen from the standpoint of the individual officer, such an education would provide insurance towards eventual retirement and a second career. Seen from the standpoint of the manpower planning departments, it would provide supernumerary personnel with something to do. Seen from the standpoint of the defense bureaucracy, it would qualify officers to deal with the supposedly much-increased scope, complexity, and ramifications of modern "defense," which, truth to tell, no longer had much to do with war at all.

Higher education as a solution to these and other problems was distinctly economic. Officers who studied required neither units to command nor equipment to run down; even their offices could be replaced by cubicles containing study desks. An entire campus, catering to hundreds of students and staff, could be funded at a cost no greater than that of a single modern air-superiority fighter.[8] Even if the professors' salaries and the purchase of a few textbooks had to be thrown in, sending officers to school was cheaper by an order of magnitude than almost anything else they could be made to do.

Once these multiple pressures had spawned the idea that officers should go to school and study, there was no lack of special pleading. Among the first to jump on the bandwagon were the universities. Early on they saw the trend simply as a new and secure source of enrollment; subsequently their horizons broadened. The first courses in "defense affairs" were taught at Harvard in the mid-fifties. Later they blossomed into a whole series of academic disciplines, ranging from "international relations" through "strategic studies" and "national security" to "peace research" and "conflict resolution." The more numerous the new security-oriented departments, centers, and institutes, and the more professors they hired, the greater the problems of funding. The greater the problems of funding, the more the universities were compelled to turn to the defense departments of the various countries, but particularly the United States, to pay their way in the form of tuition, research grants, or both.

Money, however, was but one of the issues at stake. Professors who engaged in research in the above disciplines soon began to claim expertise in fields traditionally reserved to the military and to civilian officials in ministries of defense. This put at least some universities in a position where they were able to seek and attain influence over national policy at all levels. The move from the think tanks to the councils of government originated in the United

States, where it first got under way during the Kennedy administration. It merged into the so-called "defense community," a vast interlocking complex of academics, businessmen, engineers, and managers moving in and out of government or else holding on to its tails. At the top, it has resulted in such figures as McGeorge Bundy, Walt Rostow, Henry Kissinger, James Schlesinger, Zbigniew Brzezinski, and Amitai Etzioni playing key roles in foreign and defense affairs.

There was, of course, much variation from country to country and from service to service. As late as the sixties, most senior commanders even in the United States were ex–World War II officers. Since they did not themselves hold degrees other than those awarded by their service military academies,[9] they naturally took time to see why their subordinates *did* require such degrees. Often they only let themselves be persuaded when it was pointed out to them that degrees cost money, and could therefore be used as a way to obtain funding.[10]

Other Western countries were slower off the mark. For example, when the Bundeswehr was being created during the early fifties the idea that its officers needed to be university educated did not occur to the founding fathers who, it should be noted, had themselves gone through a world war without receiving its blessings. Only during Helmut Schmidt's tenure as minister of defense from 1969 to 1973 did the Germans wake up to the fact that officers were virtually the sole leading members of society who did not hold a degree. The lack of academic accreditation made the officer's career unattractive, leading to a crisis in recruitment. Although no urgent military need existed, the armed forces universities (*Bundeswehrhochschulen*) opened their doors. The reform was successful, and has since helped create a situation in which the Bundeswehr has more young officers than it knows what to do with.[11]

The British forces, often described as the most professional in NATO, were even slower to change. Seldom in British history has the army been held in high esteem, and in fact a long tradition exists that sees the army in particular as a refuge for dunderheads; Winston Churchill was by no means the only youngster of good family considered too stupid to read law (or so he claimed) and made to enter Sandhurst instead. While this is probably no longer the case, the debate as to whether Her Majesty's Armed Forces needed officers with advanced degrees—other than specialists, that is—continued into the late seventies.[12]

Finally, the case of the Israeli Army is most instructive. The organizations that served as the forerunners of the Israel Defense Force (IDF), such as Haganah and Palmach, were founded by young people who had run away from school in order to fight for the liberation of their country. For many years afterward they continued to hold pronounced antiintellectual views, looking down on academics as butts for jokes.[13] As late as 1967 neither the minister of defense, Moshe Dayan, nor the chief of staff, Yitzhak Rabin, held academic degrees of any kind. Among fourteen officers on the general staff only one or two were university graduates, though several others had been dabbling in studies. So long as the army's social prestige stood high this fact did not cause problems and may indeed have contributed to its success.

As in the case of other modern armed forces, the IDF's awakening to the role of higher education was the result of many factors. One was the spread of such education among the civilian population, which over time made it increasingly difficult for officers to enter a second career. Another was the introduction of massive quantities of modern technology including, by most accounts, nuclear weapons; as happened in other countries also, they absolved the army of ultimate responsibility for the country's continued existence and helped change it into an organization whose main current task is policing the occupied territories. Both trends came to a head in the mid- and late seventies. This was also the time when the IDF discovered higher education—a discovery coinciding, some would say, with its incipient decline as a combat force.

As far as I know, there has been no systematic attempt to study the courses that officers of various developed countries take when they go to university. Nevertheless, the literature[14] as well as my own impressions indicate that these courses tend to fall into four main categories. First, officers have studied subjects that are perceived as somehow related to the conduct of war, such as national security, strategic studies, military theory, and, rarely, military history. The number of officers who have specialized in these subjects is, however, smaller than their intrinsic importance would suggest. The explanation is probably that officers believe, correctly, that administration, not fighting, is what modern armed forces are all about; in an age of deterrence, if they have to fight they have already failed. Then, too, military subjects as such are largely irrelevant to a second career, unless it be a career at a university or in research.

Second, the armed forces in all the advanced countries have paid much attention to the so-called social sciences, sometimes going so far as to establish military think tanks in the field.[15] Although classifications differ, this field often includes political science, "government," economics, management, business administration, sociology, communications, education, and—for those with a penchant for mathematics—operations research and systems analysis. More officers have studied these subjects, particularly business and education, than any others. Maybe this is because they are regarded as helpful not only inside the military but outside, as preparation for retirement; and, in the case of education, because it is often perceived as a "soft option."

Third, some officers have specialized in regional studies, such as Latin America, Southeast Asia, and the like. Since real-life wars are waged not against abstract "hostilians,"[16] but against flesh-and-blood opponents who are part of a given culture, such study makes excellent sense and is often encouraged by the powers that be.[17] However, serious regional study at the postgraduate level requires a thorough knowledge of languages. To many people this is a formidable barrier; another is that specialization does not normally form the highway to promotion in the military. These factors may explain why the number of officers in this group is relatively small, and also why the best intelligence services (such as the British one in World War II) are often made up largely of civilians.

Finally, many officers—particularly those commissioned into the air force or navy—go on to take advanced degrees in science or engineering. There are also many cases, however, of officers who qualify for postgraduate study in engineering or the natural sciences but prefer not to undertake it. Instead they turn to one of the above-mentioned social sciences—reasoning, one can only suppose, that managers are better paid than technicians.

There have been no systematic attempts to gauge the quality of the advanced education received by officers, either absolutely or in comparison with that of their civilian counterparts. It is important to realize, however, that most officers who study for a higher degree do so in their spare time or else during special periods provided by their services. Very often this implies attending second-rate schools, studying for shorter periods than usual, and receiving "extracurricular" instruction at the hands of second-rate instructors out to make an extra buck. Even when all this does not apply, an officer who spends a year or two taking an M.A. in such a subject as political science or economics cannot seriously hope to cross swords with the real experts either in government or at the universities. Though exceptions exist, on the whole the system violates the most elementary rules of strategy: namely, always to concentrate one's forces and never to meet one's opponent on the opponent's chosen ground.

Whether studying a variety of nonmilitary subjects at a civilian postgraduate school can do much to promote the military effectiveness of most officers is, in principle, doubtful. That staff work at the highest level demands formidable intellectual capacities is true enough, but the number of those who reach those levels is very small. On the other hand, commanding a force of naval vessels, an armored division, or a fighter-bomber wing is not primarily a question of theoretical reflection. What it does demand is a certain kind of hardheaded know-how, skill, and what the Germans call *Können,* or competence.

Although theoretical insights may reinforce one's skill up to a point, the skill itself may no more be acquired in class than studying military history makes one fit to command an armored brigade. Moreover, command is an activity that involves very heavy responsibility. It must be exercised under the extreme pressures of the battlefield, where, as Clausewitz says, even the bravest tend to behave a little strangely. Hence the cardinal qualities it demands have little to do with the talents of the civilian professional. Courage, resolution, and the ability to keep one's head are not attributes that may be acquired by sitting behind a university desk, plowing through reading lists and writing papers.

As it happened, not even the strongest advocates of postgraduate schooling were able to claim that it did much to increase the military effectiveness of its recipients. Instead, they resorted to a variety of other arguments, beginning with "the need to broaden horizons" and ending with the suggestion that officers required "integrating" with the civilian society they serve.[18] It has even been claimed, not without the indispensable statistical crutches, that the services would only be able to retain their officers by offering them a postgraduate education—which, if true, surely indicates the existence of a morale problem rather than a potential solution.

At first, officers in the United States and other countries who went to graduate school did so at their own expense and in their own free time. Though many of them probably expected study to result in improved performance and higher compensation, the military at that time had no formal method for rewarding people with advanced degrees. Later, however, the situation changed. Even as the services in general tended to shrink, the list of so-called "validated billets" allegedly requiring higher degrees expanded. Once the services began to encourage study by providing funding for it, moreover, those responsible for the budget naturally expected results for money spent. The combination of these developments meant that study began to play a role in promotion, first informally and later formally as well.

Particularly in the United States, where the prestige of the military declined after Vietnam, the movement toward postgraduate study for officers shifted into high gear during the seventies. Vast numbers of medium-ranking officers were pressured into taking advanced degrees. By and large this happened regardless of whether they were suited for study, and regardless of whether such study had anything to do with their military performance. A vicious circle was created. Officers under pressure to get their tickets punched took easy courses at less competitive schools. Easy courses at less competitive schools often produced mediocre results. Whatever the quality of the education provided, the net impact on the ability to fight and win wars was minimal. This was all the more true because the vast majority of modern armed conflicts are, as already stated, of the "low-intensity" type, which probably has less to do with university study than any other. Perhaps the entire movement is best regarded as an admission that, in an age when conventional warfare is being undercut by nuclear weapons on the one hand and low-intensity conflict on the other, military effectiveness as traditionally understood is no longer of much consequence.

The pressures that brought officers into postgraduate schools constituted only one side of the explosion in advanced military education that took place after 1945. Though the command and staff colleges in most countries date their existence to the first half of the nineteenth century, originally their mission consisted of training not senior commanders but a small number of staff (meaning desk-bound) officers who had to be more literate than the rest. In most countries, the time-honored tradition by which aristocrats in particular could reach high command without going to school was only brought to an end by World War I; to fill the gap, the system of staff colleges had to be expanded so that almost all general officers attended one on their way to the top. Even so, expansion was held within bounds. As late as 1939, the great majority not only of battalion but of regimental and brigade commanders in most armed forces never passed through a staff college, let alone a still-higher institution of military learning.

The period of demobilization after 1945 left manpower planners in armed forces throughout the world saddled with very large numbers of experienced combat officers who had attained their ranks as a result of proven competence in the field. Many of these officers had not even attended a military academy, having been commissioned from officer-training school or directly from the ranks; whatever advanced training they had received necessarily consisted of brief courses taken during the conflict itself.[19] The obvious solution was, once again, to expand the command and general staff college system. The supposed educational deficiencies may not have done much harm in war, but in peacetime they had to be corrected. At the same time the colleges acted as holding pools where surplus officers could be kept until something, somewhere, turned up for them to command.

The net result was that the share of all officers of suitable rank going to staff college was expanded very greatly during the late forties and early fifties, reaching 50 percent in the case of the U.S. Army.[20] The number of posts reserved for those who graduated also increased. Whereas the old German Kriegaksademie had only trained officers destined to serve as chief of staff *cum* operations officer of a division, in most Western armies after about 1955 all four division-level staff officers plus most aspiring battalion commanders were made to pass through school. Normally the justification for all this study was the extraordinary complexity of modern war. In practice, very often modern war had to be extraordinarily complex in order that there should be something for the officers to study. In this way most staff colleges were turned from elite institutes into mass factories, open to almost anybody who was sufficiently diligent not to be selected out during the first decade or so of commissioned service.

Over the years, manifold complicated definitions were applied to the missions of the various staff colleges. Basically, however, their task remained the same as it has always been: namely, to prepare officers for duty as commanders (though the units they commanded upon graduation were merely battalions) or else on the staffs of brigades, divisions, and corps. In practice this often meant zigzagging between training and education, theory and practice, and military and nonmilitary subjects, depending on prevailing intellectual currents and the whims of the commanding general. For example, tactical and operational training time at Leavenworth was cut by three-quarters between 1951 and 1979, when the pendulum started swinging the other way and "the operational level of war" became the most important subject taught.

Perhaps more significant than the curriculum, which varies from one college to another and tends to change every few years, are the organizational factors that most American staff colleges have in common. First, presenting a sharp contrast to the original model presented by the Kriegsakademie, and also to the current situation in most advanced countries, entrance depends on candidates' general career pattern rather than on passing an examination. This fact sets them apart from civilian postgraduate schools, where admittance is based on the students' proven academic excellence as rated by the schools themselves. It also means that the staff colleges have no say in selecting those whom they are about to instruct—a serious drawback that reflects on their prestige in the military.

Second, the duration of the courses is invariably set at one year—the U.S. Army Command and General Staff College at Fort Leavenworth being an exception in that it has recently begun selecting a small number of volunteers for a second year of study. As if to compensate, the students at most staff colleges have to spend up to twenty-five to thirty hours per week in class attending lectures, seminars, workshops, and exercises of every kind. This is far more time than at any civilian postgraduate school, and probably far more than is advisable if the goal is in-depth study.

Third, although all the colleges have exchange students from other services and pay some attention to interservice training, such training can present a problem—as became painfully obvious during the Iranian rescue attempt in 1980, when helicopters and transport aircraft belonging to different services crashed into each other. At present, the one place exclusively devoted to interservice training is the Armed Forces Staff College at Norfolk, Virginia. However, the course at Norfolk lasts for only six months. Because of its brevity, students (particularly air force officers) often select it simply as an easy substitute for their own service college. Nor is the number of students at all comparable to that of the other schools.

Fourth, although the faculty at some staff colleges rate their students in order of merit on the basis of performance and personal acquaintance, nowhere are those who take the examinations failed, or those who submit a thesis rejected; nor is it easy to see how, given the way the system is structured, a navy or air force officer the costs of whose training may run into millions of dollars should have his career aborted merely for failing to pass some school-type exams. Consequently, the atmosphere at most colleges is very different from that of a first-rate professional postgraduate school. There is often a pervasive sense of unreality, as if both students and instructors pretend to be working very hard but realize it is all make-believe.

Fifth, the faculty at most staff colleges present a problem. The military instructors tend to carry the rank of major or lieutenant colonel. They do not form part of any distinct elite, nor is it at all clear on what basis they are selected, nor do they usually stay long enough to develop real expertise as teachers. The civilian faculty, which is normally responsible for teaching subjects ranging from military history to regional studies, is in some respects even worse. Though most hold a Ph.D. and are academically better qualified than their military colleagues, they tend to be young people who failed to find a place in the general academic market. And the long hours that they spend in class do not leave them free to develop themselves in the accepted academic way, namely by doing research. To repeat, the key point is that neither the military nor the civilian faculty form part of a clearly identifiable elite such as the old German General Staff Corps. This makes it hard for their students—young officers on their way up—to take them seriously as role models, and prevents many instructors at the staff colleges from making the best use of whatever qualifications they may possess.

Finally, and possibly as a result of all these factors combined, most staff colleges in the Western world, the United States included, are not at present authorized to grant an M.A., let alone a Ph.D., in military science or in any other subject. The paradoxical outcome is that officers who want to take an advanced degree in their own profession or in a closely related one—strategic studies, say, or national security—cannot do so within the military. Instead they are referred to civilian graduate schools. There their instructors are professors who may never have seen a single day's military service. In a world that is permeated by respect for higher education—one reason why the military adopted it in the first place—the staff colleges' lack of accreditation sets them apart from civilian postgraduate schools and dooms them to second-class status. It is, indeed, tantamount to a voluntary declaration of inferiority on the part of the military as a whole.

Whereas even the youngest Western staff colleges are now close to a century old, the war colleges that stand at the apex of the military education pyramid are comparatively recent creations. We saw that World War I drove home the lesson that there was more to armed conflict than simply fighting in the field or even conducting operations in the "theater of war." Consequently, the idea of setting up additional institutes of higher military learning was "in the air" during the thirties. It was in this period that the British set up the Imperial Defense College,[21] the Germans the Wehrmachtakademie, the Americans the Industrial College of the Armed Forces, and the Soviets the General Staff Academy. Perhaps not accidentally, this was also the period when many ex–World War I officers were reaching such age and rank that they had to be either promoted or retired. The sudden simultaneous blossoming of war colleges in many countries may well have represented one way of postponing that painful choice.[22]

The trend toward adding a third tier of military training was interrupted by the outbreak of World War II. It provided the senior brass with something to do, and incidentally proved that most general officers in most armies could carry on murderously well even without the benefits of higher education. The war caused the establishment of war colleges in many countries to be postponed by a decade or so. In 1946, General George Marshall even went so far as to close down the U.S. Army War College. Six years later it was reopened, not because of any urgent military requirement but because the other services refused to close theirs.

Thus, the majority of war colleges in the East and West alike took their present form during the late forties and early fifties, with two important exceptions. For reasons connected with the legacy of World War II, the German Bundeswehr has no war college, which is not to say that its officers are the worst in NATO.[23] The

Israeli Army had one war college between 1951 and 1963 (when the superintendent was jailed as a Soviet spy) and another from 1978 on. However, the Michlalah Lebitachon Leumi (National Defense College) near Tel Aviv hardly qualifies even as a token institution of advanced learning. Its ostensible purpose is to serve as a place where senior officers are put to pasture on their way to retirement.

In Western countries, the United States included, the students who attend the war colleges tend to be lieutenant colonels and colonels in their early forties. To go to a college it is necessary to have an "outstanding" efficiency report and a recommendation from one's superior; getting one's name on the list is by far the most difficult feat that has to be performed in connection with any of the war colleges. Since the military profession is in a sense international, very often the students include visiting officers from foreign countries. Also present are civilians temporarily released from duty by government departments such as Treasury, State, and Interior in the United States—the war colleges being the only institutions of military learning that admit students who are not in uniform.

Whatever the students' formal affiliation, the war colleges' declared mission is to prepare them for performing at those elevated levels of war where military, political, economic, and technological affairs meet and merge. In doing so they have encountered problems similar to those of the staff colleges, only worse. Over the last four decades, the curricula of the war colleges have been subject to frequent changes. Often they seemed unable to decide whether they should focus on training or education, military or nonmilitary affairs, theory or practice, width or depth. The critical breaking point came in the sixties. Until then, most colleges had made a brave—albeit largely theoretical—attempt to provide the future general officer with "everything he needed to know" in the exercise of his profession. Unable to keep up with the constantly expanding definition of *defense,* however, they capitulated. Though some took longer than others, all now offered a so-called "core program" in whatever field or fields seemed most pertinent at the moment, plus an assortment of "electives" on a very broad range of subjects.

Courses at most war colleges last ten months, often including several weeks taken up by excursions in- and outside the country. This means that the time actually available for study cannot be compared to the programs of any civilian postgraduate school that takes itself seriously. The limited time probably helps explain why most war colleges prescribe classes to be attended and short papers to be written but neither major scholarly work to be performed nor academic degrees to be obtained.[24]

Whereas most countries only have a single war college, the United States, for reasons connected with history and interservice rivalry, has four, plus the Industrial College of the Armed Forces. Theoretically "National" at Fort McNair stands at the apex of the pyramid. In practice, an officer's career is not affected by the college in which he or she enrolls, and the four are interchangeable. Though the army, navy and air force war academies all have officers from sister services attending, the system is structured in such a way that there is no single place where all future American senior officers can meet and interact. Here again the United States differs from other countries, including Britain, West Germany (where the war college function is served by the Hamburg Führungsakademie), and the Soviet Union.

Since it is the mission of the war colleges to offer instruction in the higher conduct of war, the difficulties they face in balancing military and nonmilitary subjects are greater than those of the staff colleges. Particularly in the United States from about 1950 to 1975, there was a tendency to regard the higher conduct of war less as an art unto itself than as an application of the rapidly developing social sciences. Consequently the war colleges offered instruction in every conceivable "analytical" subject. On the obverse side of the coin, military history, military theory, and military arts were very often ignored.[25]

The winds of change began blowing in 1972–74 when Admiral Stansfield Turner, acting as commandant at the Naval War College, threw out the existing agglomeration of subjects as well as most electives. The curriculum was consolidated into a three-part program: (1) defense economics, focusing on the nonmilitary aspects of war as seen from the Washington, D.C., decision-making bureaucracy; (2) strategy, including its historical dimensions from Thucydides on; and (3) naval operations, with a focus on war games conducted at the college's superbly equipped war-gaming center.[26] Since then, and owing in large part to the influence of the so-called group of "reformers" in Washington, D.C., the idea that the colonels who attend the war colleges should be taught how to wage war has taken hold. This is certainly a welcome development, but it begs the question of how they became colonels in the first place.

The question as to whether the students should have to attend class or be allowed time for "free study" has also received different answers at different colleges at different times. Currently, only the Naval War College is firmly committed to the latter course: Rather than attend lectures, students spend much of their time working through reading lists that would do justice to any civilian postgraduate program. The other war colleges resemble the staff colleges in that the students spend as many as three to four hours in class each day, even before electives and special events. Education costs money, and so the normal justification behind the system is that students must be prevented from loafing at the taxpayer's expense. This argument does not carry conviction. It should be possible to trust mature men and women whose last assignment was that of battalion commander (or equivalent) and whose next assignment is that of brigade commander (or equivalent) to benefit from a year's break in their career by loafing, if necessary. If such trust is misplaced, there probably is no point in having war colleges at all.

At most war colleges, the superintendents are two- or three-star generals or their naval equivalents. Usually their appointment to the post means that they have reached the end of their career, though this did not apply to Admiral Turner, who left Newport to become commander of NATO's Southern Front and was subsequently appointed director of the CIA. This state of affairs means that their ability to attract the faculty of their choice—or, in other words, to exercise patronage—is somewhat limited. They have no influence over the selection of the students, nor can they rate graduates in a way that is significant to their careers. Hence the average superintendent can do little more than tinker with the curriculum and with his institution's internal organization.

Even less than in the staff colleges do the military faculty at the war colleges form part of any clearly identifiable military elite. Some are there because they like to teach or do research and because they believe what they are doing is important. Many, however, are where they are because they have decided to abandon the pursuit of promotion and enter a comfortable sinecure prior to retirement. Others resent their appointment, correctly realizing that it probably spells the end of any chance of being promoted to general rank. Often the faculty also includes some retired officers, either because they are qualified or because they can be paid minimal salaries. With some exceptions, they represent the system's castoffs. To hear an officer who has failed to make brigadier or equivalent expostulate about national strategy can be a painful experience in itself.

Though all the military faculty at the war colleges have M.A.'s (colonels with a Ph.D. being, for obvious reasons, rare exceptions), the same is only slightly less true of their students. Formal education therefore cannot serve as a criterion in selecting the staff; the differences between the two groups are minimal, consisting of a few years' seniority at best. At the Army War College things reached the point where not even the initiated could tell faculty and students apart, though a proposal to solve the problem by providing the former with special insignia was rejected. I personally knew a case wherein one of my students, a colonel who was more intellectually inclined than most, was appointed to the faculty of "National" at the end of the academic year. Having recovered from his initial surprise and disappointment, he was told he would teach the next class—which gave him all of six weeks to prepare himself for the new assignment.

Following Turner's reforms, the Naval War College instituted a number of highly paid Secretary of the Navy Fellowships capable of attracting top-rated personnel. For this and other reasons the civilian faculty at Newport tends to be better than most, and the college is currently the only one to have a distinguished graduate program based on serious academic work. The civilians at the remaining war colleges are paid as much, or as little, as most academic faculty in the country. They hold Ph.D.'s as a matter of course and are called professors. Nevertheless, and if only because of the long hours most of them must spend in class, few match up to their colleagues at first-class civilian institutes of higher learning.

The quality of any modern institute of higher learning is reflected by that of its publications. All the war colleges have their own heavily subsidized research centers, and all encourage both faculty and students to write. With the partial exception of the Naval War College, however, none has yet succeeded in turning itself into an outstanding center of military scholarship and doctrine.

Periodicals such as *Parameters* (the Army War College) and monographs such as the Leavenworth Series may be no worse than the average civilian publication in the field, but they are certainly no better.

More surprising still, the faculty at the war colleges—people who in theory represent the acme of military-theoretical expertise—are very often not trusted either to choose their own specialties, plan their own courses, or select their own texts. Entering the institution, they are assigned to this or that department by the powers that be. The normal procedure is for the dean to issue both teachers and students photocopied extracts from the publications of others, which they are then told to "discuss."

The most intractable problem faced by the war colleges, however, is not of their own making. It consists of the fact that, alone among all the institutes of higher learning, their intake consists of students forty years old or more. Forty is an age when one's powers to make war are, according to Napoleon, already on the wane. It is also an age when most professionals do no more than take an occasional informal seminar—if they are self-employed or work for a corporation—or else go on sabbatical, if they happen to be academics; very few of them are *required* to study by the professional association to which they belong.[27]

The students at the war colleges are officers who spend their lives operating in a highly competitive environment, have done reasonably well in their careers, and are accustomed to performing important duties. They are also at a stage in life when family responsibilities tend to be at their most demanding. By a sort of magic transformation, these people are supposed to put the clock back twenty years. They must behave like college kids, participate in seminars, write papers, and become involved in "issues."

As if all this were not difficult enough, the students are very seldom exposed to really first-class instructors except during an occasional guest-lecture basis that is largely useless. To repeat, most of the regular faculty consist of officers whose military career is at an end or else of civilians who have had difficulty placing themselves in general academe. Under such circumstances achieving anything at Western war colleges in general, and at American ones in particular, is an uphill fight; there is no lack of indications that very little is in fact achieved.[28]

We saw how, in the West, advanced military education and training exploded during the decades after 1945 and particularly from 1970 on. In a society that had less and less regard for military skills as traditionally understood, tens of thousands of officers were persuaded that training toward such skills no longer represented an adequate attainment in life. Instead, they went to graduate school, studying with civilian professors who taught them everything from pedagogy to strategy.

The forces' own system of advanced training was also expanded. The staff colleges in particular turned from elite institutes into diploma mills, attended by very large numbers of officers on their way to posts that had never before been regarded as requiring extensive formal study. Virtually all senior officers now attended war colleges, which supposedly taught them how to wage war at the medium and high levels while also sprinkling them with information about a variety of other subjects. More and more officers managed to attend both postgraduate school *and* the military colleges. In terms of years spent studying, they became perhaps the most highly qualified of any modern professional group. As explained above, however, there are many reasons why the quality of the product was less than satisfactory.

During the decades that the West became "debellicized" (to quote Michael Howard) the Soviet Union by and large followed a very different course. As a country and a people, the Soviets in 1939–1945 came much closer to destruction than did any Western society. They never forgot that military power alone saved them from this fate. The Soviets did not suscribe to the Western doctrine that from the beginning regarded nuclear weapons as a means for deterring war rather than fighting it. Not having colonies they were spared the ignominy of defeat in a "bush-fire" war, until the invasion of Afghanistan turned the tables and initiated a process whose internal political repercussions are probably still to come. Consequently the social standing both of the military and of war *per se* remained higher than in the West—an attitude that was deliberately fostered by the powers that be. There did not exist, in the Soviet Union, the feeling that officers were somehow inferior beings whose own profession was insufficient to justify their existence and who had to earn a degree from a civilian college and attend graduate school in order to redeem themselves.

Soviet perspectives on civil-military relations, too, differed from

Western ones. Aware of the fate of many past revolutions, the party elite has repeatedly worried about "Bonapartism"—a military coup—and taken steps to prevent it. On the other hand, Soviet officers, unlike their counterparts in other modern societies, have never been regarded as military technicians pure and simple. Far from being politically neutral, they are expected to act as prime carriers of communist ideology. This ideology is accordingly inculcated into them as a matter of course, and adherence to it has always played a key role in their selection and promotion procedures.[29]

The net outcome of all these factors has been that the Soviet military establishment does not believe in sending large numbers of officers to civilian universities for instruction in strategy or anything else. There is, in this sense, no such thing as advanced officer education in the USSR. A relative handful of specialists apart, what training and education they get is produced and offered by the forces themselves. Consequently the Soviet military schooling system is larger, both absolutely and relative to its civilian equivalent, than similar systems elsewhere.

The difference between the Soviet and Western system is most readily apparent at the medium level. In the United States, most junior officers will attend a short advanced-training course run by the arm to which they belong before going on to the staff college operated by their service. In the USSR, these two kinds of school are combined in a single institution. The staff academies are operated not by the services but by the arms. They function both as advanced arms schools and as staff colleges.[30] The result is a decentralized system, there being no fewer than seventeen different academies instead of three in the United States and just one in countries like West Germany and Israel. In the Soviet Union, there is a Strategic Rocket Forces Academy, a Tank Academy, an Artillery Academy, a Rear Services and Transport Academy, an Academy of Chemical Defense, a Military-Political Academy, and so on.[31]

Following the system that was originally established during tsarist days, advanced officer training in the Soviet forces starts when a lieutenant is on his first assignment a few years after he is commissioned. At this stage his superiors will "invite" him to study, meaning that he should start preparing for the entrance examinations to one of the staff academies. Lieutenants are supposed to spend two thousand to three thousand hours of their free time studying, which in two years means three hours a day, every day. How many hours' work are actually involved in preparing for the examinations is not clear; however, Albert Speer during his imprisonment at Spandau noticed how hard the Soviet jailers always seemed to study.[32]

As used to be the case in the old Kriegsakademie, responsibility for preparing the examinations rests in the hands of the academies themselves. For example, the Gagarin Air College holds a series of examinations, some written and some oral. Subjects covered include Russian language and literature; mathematics and physics; and, depending on the officer's specialty, such fields as air combat tactics, air force equipment, bombing, training, navigation, as well as command, control, and communication.[33] All this means that service as a faculty member in one of the Soviet academies carries higher prestige than that of an equivalent post in the United States, which probably makes the job of teaching easier.

Those who pass the examinations are usually captains in their late twenties. Depending on which academy they enter, they expect to study for anything between three and five years; a period during which many of them will be promoted to major. Though there is considerable variation from one academy to the next, the military subjects studied are not very different from those offered by Western staff colleges. They include operational art, general tactics, the specialized tactics of services and branches, military history, and military theory.

Soviet military academies also offer instruction to specialists. Officers can study military medicine, psychology, finance, engineering, science, and even music. The contents of all these programs are supposed to correspond to those of ordinary university courses. Consequently those who take them are allowed to become candidates for the degrees of M.A. or Ph.D. The examinations that lead to these degrees and the theses that must be written for them are read by a state-appointed committee. In addition to all this, the students receive the usual grounding in "Marxism-Leninism" that is *de rigeur* at any Soviet educational institution from kindergarten upward.

Much the most prestigious Soviet military academy remains the central one in Moscow (Frunze Academy), whose mission is to prepare officers for combined arms warfare. Its three-year courses are offered to students drawn primarily from the ground forces, with a sprinkling of officers from other services (and from foreign countries). The academy is modeled on a civilian university with "chairs" for the usual military subjects as well as military history and foreign languages. According to the *Soviet Military Encyclopaedia* (1976–) the library has some two million volumes; this is incomparably more than the holdings of any equivalent Western institute with which I am familiar.

In all the academies, most of the faculty are military men. They are medium- to high ranking officers who have themselves graduated from an academy and hold advanced degrees in military, historical or technical sciences. Unlike their Western counterparts, many of them are not assigned to their posts on a short term basis but expect to spend a considerable chunk of their career at the academy. The ranks carried by the instructors also tend to be higher than in the West; the Frunze Academy alone probably counts twenty-five to fifty general officers serving on its faculty. Much of this probably represents an inflation of high ranks and "padding." On the other hand, it undoubtedly facilitates instruction since students can look up to their instructors as successful career men who have made it in the military. According to one team of experts on the Soviet armed forces,[34] "the closest equivalent to the Frunze Academy in the US is probably the Army Command and General Staff School at Fort Leavenworth"—and yet, it is clear that there is not true equivalence in terms of faculty, curricula, library, and student-body composition.

There exists in the Soviet Armed Forces, a list—known as *Nomenklatura*—of posts that may be filled only by academy graduates. This means that, with rare exceptions, such graduates monopolize the higher ranks. It is said that, of general officers who served as military-district or as groups-of-force commanders between 1964 and 1981, 96 percent were academy graduates—and 70 percent of those had graduated from the Frunze Academy alone.[35]

Those who are not admitted to an academy, however, need not despair. Soviet military academies, like some Western staff colleges, offer correspondence courses to interested officers, most of whom are majors in their mid- and late thirties. Though it has to be carried out in the officer's free time, such study is encouraged by the forces. In time it can lead to a degree and to the reintegration of the officer, if he is still of the right age, into the mainstream of promotable personnel.

The highest institute of military learning in the USSR remains the General Staff Academy in Moscow. Like the Western war colleges, its origins date to the late thirties. World War II interrupted its development, however, and the academy as it exists today dates to 1946. The mission of this academy is defined as "preparing cadres for working in the central apparatus of the Ministry of Defense and the General Staff, in large formations and formations of all services in the Armed Forces." Students tend to carry somewhat higher ranks than their Western counterparts, counting not only lieutenant colonels and colonels but brigadier generals (one star) among their number.

The period officers spend at the Academy is two years. This is twice as long as at the American war colleges and enough for those who already hold an M.A. degree from a lower academy to study for a Ph.D. if they wish. The commandants tend to be very-high-ranking officers, for example, Marshals I. K. Bagramyon (1956–58) and M. M. Zakharov (1963–64), both of whom went on to become first deputy ministers of defense. The faculty consists of professors in uniform; between fifty and seventy-five generals and admirals are said to be assigned to the academy in various capacities. By contrast, the National War College during my period there had just one rear admiral, who was clearly at the end of his career. A general was in charge of the Industrial College of the Armed Forces, whereas the two institutes together—comprising the National Defense University—were commanded by a second.

Faculty members at the General Staff Academy invariably hold advanced degrees in military science or related subjects. Many of them also publish books and articles, conducting the kind of research that, in the United States, is carried out by the various study and analysis groups in the Pentagon and by contract research organizations. Although no strict comparison is possible, virtually all military-strategic literature published in the USSR originates in-house—that is, within the ranks of the military professionals. Such an approach is, to be sure, not without disadvantages. Any hierarchic organization may cause discussion to be stifled and the intellectual approach narrowed down. On the other hand, it has the very great merit that the Soviet military are not accustomed to looking elsewhere for expertise in their own domain—no more than physicians, say, should turn to laymen to teach them about disease.

The quality of this literature is hard to judge. First, security is much tighter in the USSR than in any Western country, with the result that authors have to be circumspect and often make their points indirectly by drawing comparisons with the forces of "Imperialist" countries. Second, the Soviet Union is not a democracy, and its armed forces are probably even less prepared to tolerate dissent than armed forces anywhere else; not everything goes. Third, many Soviet military writings bear a semipropagandistic character, a feature which sometimes makes it difficult to take them seriously. Statistical data contained in these publications are particularly suspect. They often leave the impression of having been selected at random, without even making a pretence of convincing the reader.

On the other hand, the U.S. Air Force during the last decade or so has translated a number of fundamental Soviet military textbooks. Most are excellent, and many were written by instructors at the various academies.[36] These instructors also figure prominently in the group that wrote Marshal Sokolovskii's 1962 classic, *Military Strategy.* Whatever the quality, it seems that Soviet faculty members at institutes of higher military learning do not just compile the material that they teach but write much of it; which certainly cannot be said of most of their Western counterparts.

The curriculum of the General Staff Academy has had its up and downs. For example, in 1953, Stalin's death broke the stranglehold that Stalinist military science and its "permanently operating factors" had over the development of doctrine. The new era put very great emphasis on long-range rockets and nuclear weapons as the decisive weapons in any future war. One result was that military history was dropped from the program and had to wait for twenty years before being readmitted. Thus, the Soviet system of advanced higher military education is by no means exempt from the endless reorganizations and shifts in focus that have plagued its counterparts in the West. Nevertheless, the presence of a much larger permanent military faculty probably ensures greater continuity.

The degrees awarded at Soviet institutions of higher military learning are equivalent to the corresponding ones offered by civilian universities. A military engineer with an M.A. is supposed to be the equal of a civilian engineer with the same qualifications. A doctorate in military science, such is held by many of the instructors at the General Staff Academy in particular, is the equivalent of any other doctorate. Doctors of military science who teach at the academies and are prolific writers may hope to be promoted to professor and "honored scientist," accepted into the academies of science run by the various republics, and even into the Soviet Academy of Science, which is highest-ranking institute of its kind in the Eastern bloc. All this tends to reinforce the Soviet view of the military as a profession among other professions, to be studied and mastered by the same methods and using the same criteria.

To sum up, advanced officer training and education in the present-day USSR differs from that of the West in many ways. First and most fundamental, the Soviet forces regard the problem as their own responsibility; they do not delegate any important part of it to civilian universities. Second, the use of examinations on the one hand and the awarding of degrees on the other mean that the system plays a much more important role in the process of selection and promotion. Third, the academies have some influence over who will be allowed to study and to graduate, with the result that their faculty enjoys a status much higher than that of their Western counterparts. Fourth, the courses last considerably longer than they do in the West, even if the fact that they also constitute advanced arms schools is taken into account. Fifth, they allow Soviet officers to study for degrees that are equivalent to advanced civilian degrees. Sixth, the faculty at these institutes are often much better qualified than their Western counterparts, especially when one considers the *relative* qualifications of students and teachers. This is true on paper, as far as holding advanced degrees is concerned. To judge by the fact that they produce most of their own texts, it may also be true in practice.

Unlike the Western system of staff colleges, which is relatively centralized, the Soviet one relies on a much larger group of academies that are operated by each arm separately. This arrangement has the disadvantage that it probably makes interarm and interservice cooperation more difficult; also, what the Soviets call an academy is often little more than a preliminary officer-training school where commissioned personnel are taught the fundamentals of their profession. On the other hand, the Frunze Academy and the General Staff Academy are high prestige institutions open to officers of all the arms and services. Their task is to help centralize training and ensure a uniform outlook at the highest levels. As such, they have no real rivals in the United States.

It is easy to denigrate the heavy emphasis that all the academies put on Marxism-Leninism as either nonsense or propaganda, and the accusation is undoubtedly true in many cases. Readers of Soviet military literature are familiar with Marxism-Leninism serving as cant, or else as an obligatory source of quotations that matter little one way or another. However, Marxism-Leninism may also be understood as an analytical framework. As such, its problems are neither greater nor smaller than those of any similar system. Though the dangers of a binding system of thought are obvious, this should not blind us to the possibility that an obligatory system may be better than none at all.

Finally, there should be no question in anybody's mind that Marxism-Leninism—but particularly Marxism—represents one of the great creations of the human mind. Properly used, it can offer an excellent framework for most kinds of social and economic analysis, including military analysis, both past and present. This Marxist foundation may be one reason why so much contemporary Soviet military thought appears more coherent and more consistent than anything available on the Western side. In particular, it may help explain why Soviet strategists have avoided the trap of presenting conflict as if it were merely an outgrowth of the latest technological gadget—a trap that causes most Western strategic theories to be doomed to obsolescence as soon as that gadget is replaced and, indeed, before anybody can assimilate their implications.

Whatever we think of the USSR, its regime, and its armed forces, clearly there is something to be said for a system that, putting the dangers of "militarism" aside, ensures that military professionals will be instructed primarily by military experts who carry higher ranks than do their students. The same applies to the use of the academies as instruments of promotion, to the way their faculty are selected, and to the career structure that creates a military elite by means of long years of intensive study culminating in advanced degrees. In short, the Soviet system has certain advantages that warrant serious Western consideration.

Conclusions

In attempting to identify the factors that shape the training and education of senior commanders, this study has taken the historical approach and traced the evolution of that education and that training. The story can be divided into three parts.

The first period reaches back as far as we can see and ends in 1815. Its most outstanding characteristic was that senior commanders were very seldom officers only—indeed, the very concept, as we saw, only arose during the sixteenth century. Instead, those who commanded armies were tribal leaders, politicians, statesmen, feudal barons, businessmen, great noblemen, and dynastic rulers. What they had in common was the fact that they were people whose position in society was such that, when war broke out, they *also* served as military commanders. This was true even in those cases where, as very frequently happened, it was the armed force at their disposal that had enabled them to reach those positions in the first place.

In other words, most senior commanders during this period had neither the conduct of war nor even preparation for it as their sole occupation. Rather, command was something to be exercised on a part-time, periodic basis, within the framework of their other duties or prerogatives. These, in turn, were conferred on them by their overall station in life and not merely by their capacity as soldiers.[1]

Interestingly, this applied even to such professional armies as did exist. In the Roman Army it was the rank and file who were professional soldiers and were called *miles;*[2] whereas most senior commanders entered their careers as a by-product of their family descent, social position, political standing, or a combination of these. When the Middle Ages came to an end and the first standing armies were reestablished in Europe, a similar system was followed. In the Spanish, Dutch, Swedish, and French armies between about 1450 and 1650 the rank and file consisted of paid professionals. However, their commanders were usually noblemen and grandees who owed their positions above all to their social station.

During this entire period, the only true military professionals in existence were mercenaries—that is, troops (and sometimes commanders) who moved from one war to the next, selling their services to the highest bidder. A few of these mercenary commanders, such as Iphicrates in the fourth century B.C. and some of the Italian *condottieri,* possessed educational qualifications equal to the best. However, it lies in the nature of things that neither mercenaries nor those who employed them on a more-or-less temporary basis could even think of establishing a system of formal, advanced military study.

Moreover, the overriding meaning of war was battle. To command was to command in battle. Until the end of the seventeenth century, this very often meant actual hand-to-hand fighting; witness, for example, Alexander the Great, any number of medieval commanders, and so on down to Gustavus Adolphus and Marlborough—who on more than one occasion was ridden over in cavalry melees. Under such circumstances the best training a commander could undergo probably consisted of such sports as hunting, which indeed was used for this purpose from Assyrian times right into the Renaissance and beyond.[3] The most important "theoretical" (if that is indeed the word) subject to be mastered was tactics—the art of deploying and using different kinds of troops in different kinds of terrain. Though sketches, diagrams, and later war games were sometimes employed for the purpose, then as now these problems are better studied in the field rather than in the classroom.

Both considerations together—the way one became a commander and the nature of war—explain why no formal institutions of advanced military instruction existed anywhere. This was true even at times and places that produced a substantial military-theoretical literature, such as the Hellenistic, Roman, and early modern civilizations. At most, an aspiring commander might read this literature at his leisure or else have its contents taught to him by some self-appointed instructor. However, at no time did study of this kind form a prerequisite for attaining and exercising high command.

Nowadays the medium and senior personnel of all modern armed forces undergo extensive school education. Nevertheless, even in today's technological world the view that war is the best teacher of war still holds much truth. The late Moshe Dayan knew this when, on the eve of the Suez Campaign in 1956, he recalled the Israeli officers who were studying at Camberley at the time.[4] Over the last forty years in particular, professional military have suffered any number of defeats at the hands of guerrillas and other practioners of low-intensity conflict—people who do not in the ordinary course of things undergo staff- and war-college training but have instead the authentic daily experience of combat.

The second period opened in 1815, but its beginnings were already visible in the eighteenth century which represents a period of transition. It was almost the first time in history when not only armies but a growing proportion of medium commanders came to consist of long-service personnel who were professionals and professionals only. At about the same time, the civil services of the absolutist states entered a period of sustained growth. The upshot was to free the military from such tasks as administration, police work, and tax collection—which, up to this point, had occupied as much or more of their lives as war itself. It was necessary that they be given something to do in peacetime; hence, as we saw, one reason for the opening of the first staff colleges in France and Prussia.

The rise of professional standing armies was accompanied by a substantial expansion in their size. This expansion not only enhanced the importance of administration and information processing—the functions of the general staff *par excellence*—but led to a change in the character of war as such. Strategy as understood by Jomini and Clausewitz, strategy in the sense of using battles in order to win war, was invented.[5] Particularly during the age of railroads, high command ceased to be primarily a question of leading troops in the field. Instead, much of it came to consist of mobilizing, deploying, and transporting major units, all of which could best be done with the aid of a large-scale map situated in a well-equipped office complex. Paradoxically, the task of the commander—and that of the general staff—came to an end when battle began.[6] Symbolizing the change, Moltke spent most of the 1866 campaign against Austria in Berlin. It was only during the last few days before the decisive battle of Koeniggrätz that he and his headquarters staff took the field. Even so, his share in the actual conduct of the battle was minimal. The most important thing he did was to select a cigar, with utmost care, from a case proffered by Bismarck, thus wordlessly calming his jittery superiors.

Though institutes of higher learning can trace their existence all the way back to Greece in the fourth century B.C., the profession of arms had never been included among the subjects they taught. During most of history war was regarded as a craft or an art. The idea that it also has a substantial theoretical basis that should be studied and mastered in class as a condition for exercising medium and senior command was a product of the Enlightenment. More specifically, it originated in the German *Aufklärung* with its emphasis on *Bildung* in all fields of human endeavor. It first assumed an organizational form in Scharnhorst's *Militärgesellschaft,* a debating society devoted to the propagation of military knowledge. The disastrous defeat suffered by the Prussian Army, and with it the entire *ancien régime,* at the hands of Napoleon in 1806 enabled this society to assert itself. Thereafter the Prussian Army became the first to commit itself to the principle that hardly anyone could hold a senior military post—*senior* being defined as divisional chief of staff and up—without first going through a period of intensive study.

Other European armies also set up staff colleges during the first half of the nineteenth century. Most, however, engaged in far more wars and gained much more practical experience than did the Prussians, which may be one reason why none regarded attendance at them as a very important step in an officer's career, let alone as an all-but-indispensable prerequisite for promotion to senior rank. By and large, European armed forces only began to adopt that principle after the smashing Prussian victories of 1864–1871. Anglo-Saxon ones were even slower off the mark, waiting until the turn of the century; by that time even the Bulgarians had seen the light. Nevertheless, the modern staff college *cum* general staff corps complex in many ways constituted a unique creation of the Prussian/German Army. Until 1945, there was never any question that this fact formed a critical element in its military excellence.

The success of the Kriegsakademie was due to numerous factors.

The most important ones were (1) the rigorous system for selecting students; (2) the three-year *integrated* curriculum, which for all the changes that it underwent never lost sight of the fact that its overriding function was to prepare officers for conducting war in the field; (3) the high status (and pay) enjoyed by the faculty, both in the army at large and *vis-à-vis* the students; (4) the system whereby the academy itself served as a vehicle for selection—that is, not all those who studied graduated or were taken into the general staff; (5) the preferred promotion given to graduates; and (6) the high social prestige enjoyed by the army in general, which meant that study at the academy was considered at least on a par with attendance at any civilian university.

Taken together, these factors made it very certain that study would be taken seriously. Far from being merely a token exercise, attendance at and graduation from the academy was the best thing that could happen to a young officer out to advance his career. For various reasons, some military and others social, armed forces outside Germany were never as single-minded in adopting the system. Consequently their staff colleges never became more than pale copies of the original.

To judge by results achieved between 1866 and 1945, the German system for teaching officers how to command in war has never been equaled in the modern world. Probably its greatest strength, namely the single-minded concentration on the conduct of war on the operational level, was also its greatest shortcoming: Particularly during the later decades it did not offer sufficient instruction in the nonmilitary aspects of war, including politics, economics, technology, and business administration. These were fields where the U.S. forces in particular did much better.

The third period opened in 1945. It was dominated by two factors, namely the very great expansion of civilian higher education and the shifting definition of war.

The effect of the expansion of civilian higher education was to threaten the status of the officer in society. This was all the more true because, in an age of rapid technological change and rising life expectancy, armed forces all over the world adopted the "up or out" manpower management system. Most officers were now expected to retire while still in their prime, and could expect difficulties in finding a second career.

During the same time, the impact of nuclear weapons spread

like inkstains from the major powers outward. Korea forty years ago was the last time a superpower engaged in large-scale conventional warfare, and a situation has since been created in which even less important nuclear-armed countries do not find conventional warfare a viable option.[6] In East and West alike, there has been a tendency to redefine *war* as the creation and maintenance of armed forces. In an age of deterrence, the military aspects of armed conflict are being deemphasized in favor of the nonmilitary ones.

Both developments—the spread of civilian higher education and the impact of nuclear weapons—were joined by a great many more-or-less plausible reasons to promote more advanced civilian education among officers. Among them were the need to broaden minds, integrate the military into society, prevent militarism, and keep up with technological change; nor was there a shortage of special pleading by interest groups eager to sell their services. The net outcome, starting in the sixties and (in the case of the United States) gathering steam during the seventies when the social prestige of the military reached a nadir, was a very great growth in civilian graduate and postgraduate education in the forces. The point has now been reached where virtually all American medium officers are put under pressure to take an advanced degree, since a few extra weighted points on one's efficiency report can make the difference in promotion. Other armed forces, though initially slower off the mark, are following suit.

The effect of this process on military effectiveness, by which I mean the ability to fight and win a war, has almost certainly been negligible. To judge by the case of the Israeli Army since 1982, it may even have been negative. Alternatively, the phenomenon is perhaps best taken as a tacit admission that military effectiveness as traditionally understood no longer matters much. At a time when the looming threat of nuclear escalation is more and more pushing armed conflict into the nooks and crannies of the international system, there are signs that conventional war itself may be turning into a huge exercise in make-believe.

A long peace, often coupled with a surplus of medium-ranking officers, also led to a very great expansion of the staff college system which now took in a much larger percentage of officers destined for much lower posts than formerly. The normal justification behind this growth was that war had become so much more complex; which begs the question as to why no Western army, navy or air force has retained the former three year curriculum, and why most (except the West German one) do not even offer a two year program. The real explanation is probably that peacetime armed forces in most countries no longer enjoy the social prestige that would enable them to select exceptional officers at an early age while offering a reasonable life to the rest. Unable to do this, they find themselves in a situation where their staff colleges instruct much larger numbers of officers for much shorter periods. What this means for the quality of instruction is self-evident.

Just as the importance of high schools has been downgraded by the pervasiveness of higher education, so the status of the staff colleges was eroded by the establishment or expansion of the war colleges above them. Theoretically these colleges represent the acme of military expertise; in practice they share the problems of the staff colleges, only worse. At a time when the scope of war has been greatly expanded, they offer one-year courses that often consist of a widely varied smorgasbord of different subjects. The problem of finding suitable faculty, which in turn is related to the students' age, status, and experience, is very difficult, perhaps insoluble. Nor are career-conscious officers necessarily motivated to serious study in an institution where there is no competition and from which everybody who enters is certain to graduate. Thus the best that most war colleges can offer is a year off—which may or may not be useful.

Another problem afflicting the war colleges, and most of the military system of advanced education in general, is the fact that they do not offer an opportunity to pursue an academic degree by the usual means of taking examinations and writing a thesis. In the modern world, such a deficiency is itself tantamount to an admission of second-class status; to make matters worse, an officer cannot take a degree in a subject related to his own field except by going to a civilian graduate school. The lack of accreditation represents a major reason why no Western war college has succeeded in transforming itself into a recognized center of military-intellectual excellence. It also helps explain why, when presidents and prime ministers are looking for top-level politicomilitary advice, war colleges are merely their last resort.

It has often been argued that intellectual excellence as such is not what the military are all about.[8] Their task is to produce practitioners capable of running their country's defense, not to engage in intellectual debates or issue academic publications. Few would deny that there is much truth to this argument. Still, it begs the question as to why officers on their way to the top should not spend a period during which they will be exposed to the best military intellects that a country has to offer, and why those intellects should not be concentrated at the war colleges—which, by all accounts, constitute their natural loci. Be this as it may, it seems abundantly clear that there is no point in offering any kind of instruction to ambitious, successful professionals in their thirties and forties who are on their way to the top, unless it is the very best.

Meanwhile the Soviet system of officer training and education is rather different and may in some ways be superior. For a variety reasons, some social and others doctrinal, the focus on the military aspects of war has never been lost.[9] Though the military academy–staff college system as a whole is at least as extensive as any of its Western counterparts, some of its elements remain very selective in their student intake. It was probably no accident that a group of American officers invited to visit Soviet military training institutes in 1977 were not allowed to see the Frunze and General Staff Academies—in a sense, the only ones that matter.

Overall, the program of study that leads a Soviet officer to senior command takes longer to complete than those in the West and can, unlike them, lead to academic degrees. The faculty at the general staff college outranks the students, an important advantage if the latter are to regard former as role models (something that is probably essential for good teaching). The system offers a slot for the kind of military intellectual who, in the West, would be working for a civilian university or think tank. It also carries substantial rewards, including both promotion and academic status. Finally, Marxism-Leninism, for all its dangers of rigidity and dogmatism, does offer an intellectual backbone that may be better than none and that has indeed resulted in a nuclear and conventional warfighting doctrine much more consistent than anything available in the West. For all these reasons, I believe that the Soviet system has much to recommend it. This belief will be reflected in the recommendations that follow.

Recommendations

If the way in which American senior officers are prepared for their posts is to be improved, three issues need to be addressed. First, the problem of officers' attendance at civilian universities has to be considered. Second, changes are necessary at the staff college level. Third, the war colleges must be reformed.

The present system, under which very large numbers of officers are encouraged, even pressured, into taking advanced degrees in all kinds of probable and improbable fields is, militarily speaking, quite useless. On the whole, such degrees are required only by a comparatively small number of specialists. The forces might want to have some people with advanced degrees in business administration, strategic studies, national security, military history, and the like to serve in the War Department and as instructors in staff and war colleges. When all is said and done, however, there is no question that the vast majority of officers should focus on their own profession: war. This means that they should not try to develop excessive intellectual specializations in fields other than the military one; also that they should not attempt to match government officials, businessmen, or academics in those people's areas of expertise.

On the other hand, the up-or-out career management system now in force means that most officers will have left the service by the time they reach their late forties or early fifties. Hence it is important that they be given some incentive to study for advanced degrees as a step toward retirement and an eventual second career.

The services may want to offer aid, financial or otherwise, to officers whose desire is to study; however, such study should be regarded merely as a bonus. The possession of advanced degrees is not in itself conducive to military excellence. Therefore, it should not be permitted to play any considerable role in selection and promotion procedures during an officer's active service.

To reform the staff colleges, it is necessary first of all that the age at which students enter them be lowered slightly, perhaps to twenty-eight or -nine instead of thirty-two or -three today. Entrance should not be granted to everybody but be made dependent on taking and passing an examination. Whether the examination should be "voluntary," as is the case in the USSR, or obligatory for all career officers, as in the Federal Republic of Germany, deserves careful consideration.

Once these changes are made, attendance at the colleges should be used as an active vehicle for selection and promotion, not just as a token exercise as is so often the case at present; in other words, the existing situation, under which everybody who enters graduates, should be terminated. One way of providing incentives for study would be to have the Leavenworth system of rating students adopted by all the staff colleges. The possibility of giving preferred promotion to, say, the top 10 percent of graduates also deserves consideration.

At the staff college level, the focus of study should be on military training rather than on general education. The overriding purpose of the curriculum should be to train officers to function on the staffs of divisions, corps, and armies, or in equivalent positions in the air force and navy. Students should be permitted, even required, to participate in exercises involving the highest service echelons. No one, whether army or air force or navy, should be allowed to graduate without proven mastery of the "operational level of war" at maneuvers, terrain analyses, and war games. Finally, the year spent at the college should also be used to thoroughly familiarize officers with branches of the service other than their own, this being perhaps the most important task of all.

While training at the staff college level should be predominantly practical, the possibility of extending the Leavenworth system of selecting a small group of officers for a second year of more theoretical studies deserves consideration. However, these studies should also be oriented mainly toward military topics (including study of the United States' principal projected opponents) rather than general ones. The possibility of allowing those who attend the full two-year course to take an M.A. in military science should be considered. This, of course, entails the requirement of exams or a thesis, possibly both.

To raise the prestige of the colleges and enable them to do their job, the quality of the military faculty in particular should be upgraded. Officers who teach the art of war during the first year (assuming the reform proposed above is adopted) should be experienced and *promotable* lieutenant colonels and colonels. The instructors in military history and military theory during the second year should be officers of similar rank who hold a Ph.D. On the other hand, the only civilian faculty at the staff college level should be those who teach nonmilitary subjects such as languages. The present system, whereby such subjects as military history, art and theory are often taught by young Ph.D.'s with dubious qualifications is harmful and should be abolished.

Finally, reforming the staff colleges along these lines will turn them into *military* schools. It will help provide the services with qualified personnel who are thoroughly grounded in every aspect of their profession up to and including the operational level. This, in turn, will make it unnecessary for the war colleges to teach the operational level of war to lieutenant colonels who should have it at their fingertips, thus enabling them to concentrate on their proper mission.

The present system, under which each service has its own war college, is harmful. Of the four existing service colleges, three should be closed down. Just as there is only one conduct of war at the highest level, so there should only be a single national defense university (NDU) dedicated to studying and teaching it. Such a step would do away with the duplication that now exists between the medium- and senior-service schools, particularly in the navy. In addition, it is in the spirit of the 1986 Defense Reorganization Act.

The question of the location the new NDU deserves careful thought. It should not be in Washington, D.C., which, as the center of power and news, in many ways offers too vulnerable an environment for serious thought and study. Nor, on the other hand, should it be located in some provincial backwater where a first-rate faculty will be impossible to assemble and retain. These considerations indicate that, among the existing locations, Newport, Rhode Island, is almost certainly the most suitable, with Carlisle Barracks, Pennsylvania, as second choice.

Lowering the age at which officers enter the staff colleges, as proposed above, will make it possible to carry out a corresponding reform at the war college level. Students should enter the reformed National Defense University at ages thirty-eight or -nine (preferably even younger) instead of forty-one or -two as they do now. They should be hotshot officers on their way to the top, not middle-aged bureaucrats in uniform, as tends to be the case at present.

Since officers who enter will already have the operational level of war at their fingertips, the university will be able to focus on its proper mission, which is to prepare them for exercising command at the highest levels. This mission, in turn, could be divided into three parts. First, the university should emphasize joint operations—that is, operations involving the cooperation of more than one service. This part of the program should include joint exercises, maneuvers, terrain analyses, war games, and extensive visits to the bases and facilities of sister services.

Second, the curriculum should stress the nonmilitary—that is, the political, social, and economic—aspects of war, including the structure of industry, mobilization, finance, and public relations and also including corresponding problems as they affect the United States' main opponents. It is important that these subjects be taught by the best available faculty in the field, including both academics and ex-government officials. Instructors should be well known in their fields, highly paid, and older and more experienced than those they teach.

Third, the university should encourage the students to enter one nonmilitary subject completely unconnected with their specialty. The objective of this is to enable them to leap over their own shadow (that is, get away from what they are and acquire a perspective quite different from the one to which their careers have accustomed them). Even more than the nonmilitary aspects of war, these subjects—Chinese, say, or Greek philosophy—ought to be taught by the very best faculty available.

Assuming this modified program is too large for a single year, the extension of the course to two years should be considered.

The majority of officers, studying for one year only, will focus mainly on joint operations. The elite who, on the basis of their performance at the university itself, are invited to stay a second year should study the nonmilitary aspects of war as well as one field with which they are completely unfamiliar. Those who attend the two-year course should be allowed to take a Ph.D. in national defense or strategic studies—which again entails the requirement of examinations, a thesis, or quite probably both.

Since the nonmilitary aspects of war include the problems of industrial organization and mobilization, the extension of the course to two years will permit the Industrial College of the Armed Forces to be closed down. This presents the additional advantage that all American officers above the rank of colonel, regardless of their service affiliation, will proceed through a uniform training system.

In so far as no school can be better than its faculty, no other factor is so vital to reform as a drastic upgrading of the staff. It is imperative that the system whereby an appointment to the faculty of a war college usually spells the end of an officer's military career be terminated. Instead, the German system, under which instructors at the Kriegsakademie were the intellectual flower of the forces and the envy of their comrades, should be adopted.

Simultaneously, the quality of the civilian faculty should also be drastically upgraded from the present system, which is not competitive with the better academic institutions. To this purpose they should be given greatly increased financial rewards. Instructors should, with rare exceptions, be older and more experienced than their students; this is all the more true because those students are not mere college kids but consist of highly experienced, highly motivated officers.

In order to help attract a first-rate faculty, and to upgrade the quality of instruction, the number of hours they spend in class should be sharply cut until it corresponds more closely to those devoted to teaching at civilian universities. The faculty should also be made to engage in original research—this being the only way to ensure that, in a world characterized by rapid change, they will remain abreast of developments with their specialties.

To further upgrade the quality of both study and teaching at the reformed national defense university, the number of hours students spend in class should be sharply reduced and made comparable to that decreed by a first-class civilian university for its graduate school. It must be possible to trust students who have served as battalion commanders or in equivalent positions (and whose next assignment is that of brigade commander) not to loaf. If such trust is misplaced, then probably there is no point in having war colleges at all.

To make sure they do not loaf, the present system, under which anyone who enters a war college graduates from it, should be terminated. In the case of officers who are around forty years old and hold the rank of lieutenant colonel, negative incentives such as examinations or rating make little sense. Probably the strongest incentives that can be offered consist of being accepted into the second-year course and being allowed to take a Ph.D. A Ph.D. in military science or a related subject taken at the reformed NDU, as outlined above, *should* count as a factor in the promotion game.

As will be seen readily enough, the above recommendations are all connected to each other. Getting rid of superfluous advanced degrees in civilian fields will enable officers to concentrate on their profession. Concentrating on their profession, they will be able to make faster progress through the hierarchy of rank, entering the staff and war colleges at an earlier age. Lowering the age at which officers become eligible for admission presents the additional advantage that it will automatically increase the number of candidates, thus acting as a spur to competition. Competition, so long as it is not driven to cutthroat extremes, should help improve both the quality of study and that of commanders in general.

Besides enabling officers to study at a period in their lives that is most conducive to it, lowering the students' age will also make it easier to find first-rate faculty (both military and civilian) who are older and more experienced and *to whom the students can look up*. Such faculty, in turn, should be given a say in the rating and promotion of officers, thus increasing their authority. All of this is tantamount to saying that selection for high command should take place at an earlier stage in a man's career; also, that study, provided it is of the right kind, *should* play a role in the promotion game.

These advantages apart, the system here proposed will improve interservice coordination, cut away unnecessary educational flab, and, by closing down some existing programs and schools, lead to substantial economies. Every part of it is linked to every other; piecemeal reforms will be to no avail.

Notes

Chapter 1
Parameters

1. See Crackel, "The Founding of West Point"; particularly p. 534 ff.
2. On the Israeli system of officer selection and training, see Lester, "Israeli Military Psychology"; also Gal, *A Portrait of the Israeli Soldier,* chap. 7. Gal, incidentally, is a retired colonel whose last assignment was as chief of military psychology, IDF.
3. On the military academy system, see Forman, *West Point; a History of USMA;* Ambrose, *Duty, Honor, Country;* as well as Ellis and Moore, *School for Soldiers.* On the origins of the Naval Academy see Karsten, *The Naval Aristocracy.* Though far larger numbers of officers are commissioned from ROTC programs than from the military academies, the ROTC has not attracted its fair share of publications; see Lyons and Masland, *Education and Military Leadership.*
4. On the way the Soviets handle these questions, see Scott and Scott, *The Armed Forces of the USSR,* p. 335 ff.
5. On the West German system of officer training see van Creveld, "Bundeswehr Manpower Management," p. 58 ff; also, in greater detail, Radbruch, "From Scharnhorst to Schmidt."
6. While researching my book *Fighting Power* in the archives of the German Army at Freiburg in 1980, I came across a 1932 Reichswehr study—whose reference number I unfortunately no longer have—which argued that the best officer material consisted not of college graduates but of youngsters with no more than a high school education. An Israeli study, mentioned in Gal, p. 35, arrived at similar conclusions.

7. For the Soviet handling of the officer's career see Scott and Scott, *Armed Forces of the USSR,* chaps. 10 and 11; also Erickson, "Soviet Military Manpower Policies."
8. For example, the Red Army, the North Vietnamese Army, and the Palmach, forerunner of the Israeli Army prior to the establishment of the state. The German Wehrmacht also refused to separate politics from leadership, but this is hardly a commendable example.
9. The modern military's attitude to their political masters is often compared to that of a professional (a lawyer, say, or a doctor) to their clients; see Millet, *Military Professionalism and Officership in America.* It is important to realize, however, that such has by no means always been the prevailing view. During much of history the functions of political rule and military command were united on every level from the top down, and some of the world's greatest rulers from Julius Caesar on owed their position primarily to their military prowess.
10. On the problem of surplus officers in the U.S. military see Luttwak, *The Pentagon and the Art of War,* chaps. 6 and 7.
11. See Philip, *Étude sur le service d'état major,* p. 5 ff.

Chapter 2
Origins

1. See, for example, 2 Sam. 18:10–17, 19:5–7.
2. Plutarch, *Pericles,* iv.
3. Ibid., *Alcibiades,* ii, 2; vii, 2.
4. For classical Greek education in general, see Jaeger, *Paideia,* particularly vols. 1, pp. 3–15, and 2, pp. 251–58.
5. On the ephebate see Vidal-Naquet, "Le chasseur noir et l'origine de l'éphèbie athénienne." On military Greek military education in general, Plato, *Laches,* 179–84.
6. For a summary of what little is known of Alexander's education, including his military education, see P. Green, *Alexander of Macedon,* p. 54 ff.
7. The workings of the Greek system of *xenia,* or host-relatives, has been described by Herman, *Guest-Friendship in Ancient Greece.* The idea that it may have served to train youths in war stems from him.
8. Plutarch, *Philopoemen,* iv, 4.
9. Ibid., iv, 5.
10. Ibid., *Flamininus,* i, 3.
11. The best short discussion is Saller, *Personal Patronage under the Early Empire,* chap. 3.
12. Caesar, *De Bello Galico,* book 1, chap. 39.

13. Nothing specific has been written on the training of senior Roman commanders. See, however, Dobson, "The Rangordnung of the Roman Army"; and, in general, Bonner, *Education in Ancient Rome.*
14. A textbook example of the career of one knight who rose to become a very senior commander indeed is contained in Duby, *Guillaume le maréchal.* See also McRobbie, "The Concept of Advancement in the Fourteenth Century."
15. See Wisman, "L'Epitoma rei militaris."
16. See Allmand, *Society at War,* particularly chap. 2.
17. Much the best modern account of the workings of a mercenary army, including its officers, is Parker, *The Army of Flanders and the Spanish Road.* See also Parker's *The Military Revolution.*
18. Little, to my knowledge, has been written on the origins of most of these academies. For the curricula of the French ones, at any rate, see Artz, *The Development of Technical Education in France,* which has several chapters on the subject.
19. Montecucculi's reading list has come down to us. It included, besides military works both ancient and modern, books by the foremost philosophers and scientists of his age. See Gat, "Montecucculi: Humanist Philosophy."
20. This was Thiebault, *Manuel des adjutants généreaux.* Much the best recent analysis of eighteenth-century military thought is contained in Gat, *From the Enlightenment to Clausewitz.*
21. Fredericus Rex, *Werke,* vol. 4, chap. 18. For the development of the quartermaster staff see also Creveld, *Command in War,* p. 35 ff; for the work of a typical quartermaster general (Marlborough's William Cadogan), Chandler, *The Art of War in the Age of Marlborough.*
22. See Goerlitz, *History of the German General Staff,* p. 5; and Duffy, *The Army of Frederick the Great* pp. 37–39.
23. Philip, *Étude sur le service, d'état major,* p. 6.

Chapter 3
Comparisons

1. For one eighteenth-century catalog of this kind see Feuquières, vol. ii, p. 364ff. For another, Turpin de Crise, *An Essay on the Art of War,* p. 268ff.
2. The first military writer to use the term *strategy* in anything like its modern sense was, it seems, Joly de Maizeroy in *Cours de tactique.*
3. For the way battles were commanded before 1800, see van Creveld, *Command in War,* pp. 41–55.

4. Duffy, *The Army of Frederick the Great,* p. 37.
5. For a short biographical sketch in English see Paret, *Clausewitz and the State,* p. 59ff.
6. On Scharnhorst's reorganization of the academy, see Klippel, *Das Leben des Generals von Scharnhorst,* vol. 3, pp. 237–55.
7. Blumentritt, "Militärisches Schulsystem und Hochschule," p. 670.
8. Ibid.
9. On Peucker's reforms see Bald, *Bildung und Militär,* p. 67.
10. Comparative statistics reveal that the Kriegsakademie was much superior, as a ladder for social mobility, to the universities. See Knight, *The German Executive,* pp. 36, 41, 45.
11. For the examinations see Guderian, "Zur Geschichte des deutschen Generalstabes," pp. 2–3. This was one of the studies undertaken for the chief of military history, U.S. Army, after World War II.
12. See Erfurth, *Die Geschichte des deutschen Generalstabes,* pp. 139–40.
13. Model, *Der deutsche Generalstabsoffizier,* p. 38.
14. Reinhardt, "Training of Senior Officers," p. 3.
15. On the Reinhardt Courses see Bald et al., *Tradition und Reform im Militärischen Bildungswesen,* pp. 167–68.
16. The best account of Hitler's intellectual world is probably Maser, *Hitler,* chaps. 5, 9. See also van Creveld, "Warlord Hitler—Some Points Reconsidered."
17. An excellent case in point took place in July 1941 when Hitler, arguing that "my generals know nothing about the economic side of war," overruled the general staff and sent the Panzers into the Ukraine rather than straight toward Moscow. See Guderian, *Panzer Leader,* p. 200. I have examined this problem at greater length in my article, "On Learning from the Wehrmacht and Other Things."
18. On these officers see van Creveld, *Command in War,* pp. 67, 75–78.
19. For details see Chalmin, *L'officier français de 1815 à 1870,* p. 295ff.
20. A new German edition of Vauban was published in 1830, the year that the French besieged Antwerp during Belgium's war of independence.
21. Chalmin, *L'officier français de 1815 à 1870,* p. 315.
22. Ibid.
23. Thiebault, *Mémoires,* vol. 5, p. 411.
24. See Howard, *The Franco-Prussian War, passim.*
25. Riff, *Histoire de l'Ex Corps d'État Major,* pp. 182–83.
26. Stoffel, *Rapports militaires,* pp. 27, 111, 112, 120, 124–28.

27. Revol, "A l'École Supérieure de Guerre," 1903–1905," 132.
28. For figures see Kuntz, *L'officier français dans la nation,* p. 123.
29. See Dutailly, *Les Problèmes de l'Armée,* p. 242.
30. Doughty, *The Seeds of Disaster,* p. 76.
31. Ibid.
32. Dutailly, *Les Problèmes de l'Armée,* pp. 240–42.
33. Quoted in Bond, *The Victorian Army and the Staff College,* p. 78 n. 8.
34. Godwin-Austen, *The Staff and the Staff College,* p. 28.
35. Bond, *The Victorian Army and the Staff College,* p. 55.
36. E. Spiers, *The Army and Society 1815–1914,* p. 151.
37. On the first and last of these see Luvaas, *The Education of an Army,* chaps. 5, 7; also Godwin-Austen, *The Staff and the Staff College,* pp. 133–34, 231–32. Both Hamley and Henderson, incidentally, were valued for their liking of games.
38. Bond, *The Victorian Army and the Staff College,* p. 62.
39. The present West German system for dealing with it is to pass all officers through a four-month *Grundlerhrgang* (basic course) prior to the examination.
40. See F. W. Young, *The Story of the Staff College,* p. 8.
41. On this problem see Barnett, *Britain and her Army,* p. 344ff.
42. Godwin-Austen, *The Staff and the Staff College,* p. 157.
43. One finds irresistible the following quotation from an admiring contemporary biography of Field Marshal Sir John French: "Inevitably, games quickly took possession of [the young John French's] imagination. Very soon the war game had first place in his affections" (Chisholm, *Sir John French,* p. 3).
44. P. Parker, *The Old Lie—The Great War and the Public School Ethos.* See also Veith, "Play Up! Play Up! and Win the War!" Though England was not the only contemporary society which looked at war as a game, nowhere else was an entire educational system based on that idea.
45. Quoted in Bond, *The Victorian Army and the Staff College,* p. 329.
46. Before 1900 the navy only had a Royal Naval College which taught sub-lieutenants technical subjects. In that year four months courses on naval history, strategy, tactics and international law were established and made compulsory for senior officers. See Marder, *From the Dreadnought to Scapa Flow,* vol. 1, pp. 32–33.
47. General Sir H. E. Franklyn, quoted in Young, *The Story of the Staff College,* p. 25.

48. See Bond, *British Military Policy between the Two World Wars,* p. 37.
49. This figure, like most of the following account, comes from Kulikov, *The General Staff Academy.* Though an English summary of this study was published in ms. form in 1982 it leaves out many important details. Thanks are due to my student, Mr. G. Koren of the Israeli Foreign Office, for helping me translate and exploit this and other Soviet sources.
50. For a disparaging view of these officers see Seaton and Seaton, *The Soviet Army,* p. 77ff.
51. In fact, NATO in Europe both outspends the Soviets and maintains more manpower under arms. Soviet superiority consists mainly in organization, which allows the maintenance of far more heavy weapons per unit.
52. Even Seaton, before he turned anti-Soviet, wrote: "The Soviet High Command did make errors, some of them grievous ones, and continued to do so throughout much of the war, and many of these were probably due to Stalin's dominance of the *Stavka.* Yet in spite of these mistakes, the war direction of the GKO and the *Stavka* was in many ways superior to that of the German OKW and OKH" (*The Russo-German War,* p. 85).
53. Upton, *The Armies of Asia and Europe.*
54. Quoted in McGregor, "The Leavenworth Story," pp. 70–71.
55. Quoted in Hattendorf et al., *Sailors and Scholars,* p. 18.
56. The British Navy at the time had only higher technical instruction, not a war college. The institution of the latter had to wait until the Fisher era, 1905–6, when it was carried through against much opposition.
57. See Pratt, *The United States Naval War College,* appendix C.
58. For an account of this period see Dastrupp, *The U.S. Army Command and General Staff College,* pp. 60–77.
59. Ball, *Of Responsible Command,* p. 182. Leavenworth was said to produce "mental indigestion."
60. See Masland and Radway, *Soldiers and Scholars,* pp. 81, 84–85.
61. *The Brain of an Army,* by British military historian and amateur soldier Spenser Wilkinson, an admirer of the German general staff. Root was also influenced by Upton's *The Military Policy of the United States*—a strong critique of American military policy—and even had it published as a government paper.
62. See Pappas, *Prudens Futuri,* p. 31ff.

63. Masland and Radway, *Soldiers and Scholars,* p. 97.
64. General Order No. 7, War Department, Washington, D.C., February 25, 1924.
65. See "Report and Recommendations of a Board," p. 125.
66. See Euliss, "War Gaming at the U.S. Naval War College"; also Vlahos, "Wargaming, an Enforcer of Strategic Realism."
67. See van Creveld, *Fighting Power,* chap. 11.

Chapter 4
Problems

1. The first to propose that officers should be university educated was the French socialist Jean Jaurès in *L'armée nouvelle.* His purpose was "to break with the aristocratic and monk-like habits of the military academies"—scarcely an idea to recommend itself to the armies of the time.
2. Menninger, *Psychiatry in a Troubled World,* p. 110.
3. This and subsequent figures from Harris, *A Statistical Portrait of Higher Education,* pp. 294, 433, 930. The figure of 30,000 Ph.D.'s for 1970 is from Pusey, *American Higher Education 1945–1970,* p. 55.
4. See Table 9–3 in Bowen and Schuster, *American Professors,* p. 179.
5. For example, the Israeli military took no interest whatsoever in the half-mile of beach known as Taba, which formed the subject of a nine-year dispute between Israel and Egypt. That is the kind of problem they have been content to leave entirely to the discretion of the foreign ministry.
6. See for example Fuller, *The Conduct of War, 1789–1961;* and Falls, *The Art of War.* On the Soviet side, the belief in the total character of the next war was reasserted when V. D. Sokolovskii brought out his authoritative *Military Strategy* in 1962.
7. See Luttwak, *The Pentagon and the Art of War,* chaps. 6, 7, and appendixes.
8. The $36 million Congress appropriated in 1987 to consolidate the National Defense University campus in Washington, D.C., is just about enough to buy an F-15 air-superiority fighter with spares.
9. See Margiotta, "A Military Elite in Transition," Table no. 1.
10. As early as 1973, spending for officer education in the United States was running at $130 million a year; Sarkesian and Taylor, "The Case for Civilian Graduate Education," p. 251. It may not be coinci-

dental that the greatest expansion of officer education took place in the years of fiscal retrenchment following Vietnam.

11. On these reforms see Radbruch, "From Scharnhorst to Schmidt"; also in greater detail Libau, *Akademiker in Uniform.*
12. Hartley, "Science Graduates in the Army,"
13. For example, as late as 1970 only 25 percent of all career officers (comprising a fraction of all officers) had a university degree of any kind; Luttwak and Horowitz, *The Israeli Army,* p. 182. When I taught at the Command and General Staff School in 1985–86 I was told that only 40 percent of my students even had a high school diploma.
14. See Janowitz, *The Professional Soldier;* also Lyons and Morton, *Schools for Strategy.*
15. For example, the Bundeswehr's *Sozialwissenschaftlichesforschung-samt* in Munich.
16. P. Dickson, *The Electronic Battlefield,* Bloomington, Ind., 1976.
17. For example, the Pentagon offers no-strings-attached scholarships to those who agree to specialize in Arabic studies.
18. Sarkesian and Taylor, "The Case for Civilian Graduate Education," Taylor and Bletz, "The Case for Officer Graduate Education."
19. For general staff training during the war see Model, *Der deutsche Generalstabsoffizier,* pp. 111–43; Masland and Radway, *Soldiers and Scholars,* pp. 100–104.
20. Masland and Radway, ibid., p. 281; Hart and Lind, *America Can Win,* p. 168.
21. See Chegwidden, "The Imperial Defence College."
22. See Bond, *British Military Policy,* p. 50, for the game of musical chairs then played by senior officers; a similar situation prevailed in France.
23. On the present-day West German system of training officers see Majewski and Peyton, "German Army General Staff Officer Training," pp. 23–24.
24. At present, only Leavenworth is accredited to award an M.A. in "military science." Over the last quarter-century there have been many other attempts to accredit the war colleges, but all ended in failure when it was found that they just would not meet the requirements of civilian universities.
25. One result of this was that the old translation of Clausewitz's *On War,* originally done by Colonel J. J. Graham in 1898, continued in service. It was only in 1976 that it was finally replaced by the one by M. Howard and P. Paret.

26. For details see Murray, "Grading the War Colleges," *The National Interest;* also Hart and Lind, *America Can Win,* pp. 171–72; also F. H. Hartmann, "The Naval War College in Transition."
27. On the way other professions organize these things see Houle, *Continuing Learning in the Professions.* From this, too, it appears that the military is unique.
28. See Brewer, "The Impact of Advanced Education." To judge by this article, the most important impact is a move toward increased "pragmatism"—a curious outcome from a year of largely theoretical studies.
29. See Odom, "Bolshevik Ideas on the Military Role."
30. The Malinovsky Tank Academy, for example, offers hands-on instruction in driving, firing, and repairing tanks, as well as in armored warfare doctrine. See Head, "Soviet Military Education."
31. For a list and short description see Scott and Scott., *The Armed Forces of the USSR,* pp. 356–57.
32. Speer, *Spandau,* p. 267.
33. Scott and Scott, *The Armed Forces of the USSR,* p. 352.
34. Ibid., p. 357.
35. Jones, *Red Army and Society,* p. 91.
36. For example, Lomov, *The Military-Scientific Revolution,* and Druzhinin, *Concept, Algorithm, Decision,* probably the best work on the subject in any language.

Chapter 5
Conclusions

1. The term *soldier* was coined in the sixteenth century when it meant "somebody who receives sold [pay]." Thus, at first, it did not encompass all the members of an army but only those of them who received wages, which by and large meant the rank and file.
2. The Middle Ages translated *miles* as "knight." A knight, however, was not a professional and certainly did not go to school to study war.
3. See, for example, Machiavelli, *The Prince,* p. 88.
4. Dayan, *Diary of the Sinai Campaign,* p. 37.
5. Creveld, *Command in War,* chap. 2.
6. In 1973, fear of a possible nuclear holocaust probably helped limit the Egyptian and Syrian offensives to a depth of a few kilometers. In 1982 the Falklands War could take place only because nobody knew where the Falklands were.

7. On this problem see Ludendorff, *My War Memoirs 1914–1918,* vol. 2, p. 298 ff.
8. Most recently and cogently by Ball, *Of Responsible Command,* pp. 491–92.
9. Gorbachev's announcement of a new military doctrine in 1988, insofar as it is not pure cant, may change this situation.

Bibliography

Abbreviations used in the Bibliography:
AF&S Armed Forces and Society
JCH Journal of Contemporary History

General

Andreski, S. *Military Organization and Society.* Berkeley, Calif., 1968.

Barnett, C. "The Education of Military Elites." *JCH* (July 1967): 15–35.

Baudissin, W. von. *Officer Education and the Officer's Career.* Adelphi Paper No. 103, London 1973.

Creveld, M. van. *Command in War.* Cambridge, Mass., 1985.

———. *Fighting Power.* Westport, Conn., 1982.

Doorn, J. van, ed. *Military Profession and Military Regimes.* The Hague, 1969.

Falls, B. *The Art of War.* New York, 1961.

Fuller, J. F. C. *The Conduct of War, 1789–1961.* London, 1962.

Hittle, J. D., *The General Staff, its History and Development.* Harrisburg, Pa., 1961.

Little, R., ed. *Handbook of Military Institutions.* London, 1971.

Lang, K. *Military Institutions and the Sociology of War.* London 1972.

Wilkinson, S. *The Brain of an Army.* London, 1890.

Premodern

Allmand, C. T. *Society at War, the Experience of England and France during the Hundred Years War.* Edinburgh, 1973.

Bibliography

Angelin, J. P. "The Schools of Defense in Elizabethan London." *Renaissance Quarterly* (Autumn 1984): 393–410.

Artz, F. B. *The Development of Technical Education in France 1800–1850.* Cambridge, Mass., 1966.

Barker, T. M. *The Military Intellectual and Battle. Raimondo Montecuccoli and the Thirty Years War.* Albany, N.Y., 1975.

Bonner, S. F. *Education in Ancient Rome.* London, 1977.

Caesar, G. J. *De bello galico.* London: Loeb Classical Library, 1946.

Chandler, B. *The Art of War in the Age of Marlborough.* London, 1976.

Childs, J. *Armies and Warfare in Europe 1648–1789.* New York, N.Y., 1982.

Dobson, B. "The Rangordnung of the Roman Army." *Actes du septième congrès international d'epigraphie.* Paris, 1979. Pp. 191–204.

Duby, G. *Guillaume le maréchal, ou le meilleur chevalier du monde.* Paris, 1984.

Duffy, C. *The Army of Frederick the Great.* London, 1974.

Fredericus Rex. *Werke.* Vol. 4, *Militärische Schriften.* G. B. Holtz ed. Berlin, 1913.

Feuquières, J. de. *Mèmoires Historical and Military.* Westport, Conn., 1968; original French ed., Paris, 1736, vol. 2.

Gat, A., *From the Enlightenment to Clausewitz.* Oxford, 1988.

———. "Montecucculi: Humanist Philosophy, Paracelsian Science and Military Theory." *War and Society* (Sept. 1986): 21–31.

Green, P. *Alexander of Macedon.* London, 1976.

Herman, G. *Guest-Friendship in Ancient Greece.* Cambridge, 1988.

Jaeger, W. *Paideia.* Oxford, 1946.

Joly de Maizevoy, P. G. *Cours de tactique.* Paris, 1785.

Kuppel, G. H. *Das Leben des Generals von Scharnhorst.* Leipzig, 1869–71.

Machiavelli, N. *The Prince.* London: Penguin, 1961.

McRobbie, K. "The Concept of Advancement in the Fourteenth Century in the Chroniques of Jean Froissart." *Canadian Journal of History* 6 (1971): 1–9.

Paret, P. *Clausewitz and the State.* Princeton, N.J., 1976.

Parker, G. *The Army of Flanders and the Spanish Road.* London, 1972.

———. *The Military Revolution.* Cambridge, 1988.

Plato. *Laches.* Loeb Classical Library ed. London, 1952.

Plutarch, *Lives.* Loeb Classical Library ed. London, 1950.

Riff, E. A. *Histoire de l'Ex Corps d'État Major* Paris, 1881.

Saller, R. *Personal Patronage under the Early Empire.* Cambridge, 1982.

Thiebault, P. *Manual des adjutants generaux et des adjoints employes das les états majors des armées.* Paris, 1806.

Turpin de Crise, L. de. *An Essay on the Art of War.* London, 1761.

Vattel, E. de. *The Law of Nations.* Philadelphia, 1853.

Vidal-Naquet, P. "Le chasseur noir et l'origine de l'éphèbie athénienne." *Annales Économiques, Sociales, Culturelles* (1968): 947–64.

Wisman, J. A. "L'Epitoma rei militaris de Vegece et sa fortune au moyen age." *Le Moyen Age* 85 (1979): 13–31.

Upton, E. *The Armies of Asia and Europe.* Washington, D.C., 1878.

Modern Germany

Bald, D. *Bildung und Militär, das Konzept des Reformers Eduard von Peucker.* Munich 1977.

Bald, D., et al., eds. *Tradition und Reform in Militärsichen Bildungswesen.* Baden Baden, 1985.

Blumentritt, G. "Militärisches Schulsystem und Hochschule," *Wehrkunde* (Dec. 1959).

Creveld, M. van. "Bundeswehr Manpower Management." *RUSI and Brassey's Defence Yearbook 1983.* Oxford, England, 1983, pp. 47–72.

———. "On Learning from the Wehrmacht and Other Things." *Military Review* (January 1988): 63–71.

———. "Warlord Hitler—Some Points Reconsidered." *European Studies Review* 4 (1974): 57–79.

Demeter, K. *The German Officer Corps in Society and State.* London, 1965.

Dupuy, T. N. *A Genius for War.* London, 1977.

Erfurth, W. *Die Geschichte des deutschen Generalstab von 1918 bis 1945.* Göttingen, 1957.

Goerlitz, W. *History of the German General Staff, 1657–1945.* New York, 1953.

Guderian, H. "Zur Geschichle des deutschen Generalstabes—November 1948." Study prepared for Militärgeschichtliches Forschungsamt, Freiburg im Breisgau.

———. *Panzer Leader.* London, 1952.

Herwig, H. H. *The German Naval Officer Corps: a Social and Political History.* Oxford, England, 1973.

Kitchen, M. *A Military History of Germany.* London, 1975.

Knight, E. M. *The German Executive 1890–1933.* Stanford, Cal., 1933.

Libau, W. E. *Akademiker in Uniform—Hochschulreform in Militär und Gesellschaft.* Hamburg, 1976.

Ludendorff, E. *My War Memoirs 1914–1918.* London, n.d.

Majewski, N. and J. H. Peyton. "German Army General Staff Officer Training." *Military Review* (December 1984): 23–34.

Maser, W. *Hitler.* London, 1973.

Meier-Welcker, H. *Untersuchungen zur Geschichte des Offizier-Korps.* Stuttgart, 1962.

Model, H. G. *Der deutsche Generalstabsoffizier.* Frankfurt am Main, 1968.

Radbruch, H. E. "From Scharnhorst to Schmidt: The System of Education and Training in the German Bundeswehr." *Armed Forces and Society* 5 (1978–79): 606–25.

Reinhardt, H. "Training of Senior Officers," manuscript prepared for U.S. Army, Historical Division, European Command, Königstein im Taunus, 1951.

Speer, A. *Spandau, the Secret Diary.* London, 1976.

Wagemann, E. "Das Studium im Rahmen der Ausbildung zur Offizier des Truppendienst." *Europäische Wehrkunde* 29 (1980): 41–44.

Modern France

Chalmin, P. *L'officier francais 1815–1870.* Paris, 1957.

Chuquet, A. *L'école de Mars.* Paris, 1899.

Doughty, R. A. *The Seeds of Disaster—the Development of French Army Doctrine 1919–1939.* Hamden, Conn., 1985.

Dutailly, H. *Les Problèmes de l'armée de terre française, 1935–1939.* Paris 1980.

Holmes, R. *The Road to Sedan.* London, 1984.

Howard, M., *The Franco-Prussian War.* London, 1961.

Jaurès, J. *L'armée nouvelle.* Paris, 1911.

Kuntz, F. *L'officier français dans la nation.* Paris, 1960.

Monteilheit, J. *Les Institutions militaires de la France, 1814–1924.* Paris, 1926.

Philip, A. de. *Étude sur le service d'état major pendant les guerres du premier Empire.* Paris, 1900.

Revol, J. "A l'École Supérieure de Guerre," *Revue Historique de l'Armée* 3 (1979): 130–44.

Stoffel, E. G. H. *Rapports militaires 1866–1870.* Paris, 1871.

Thiebault, P. de. *Mémoires.* vol. 5. Paris 1897.

Modern Britain

Bond, B. *British Military Policy Between the Two World Wars.* Oxford, England, 1980.

———. *The Victorian Army and the Staff College, 1854–1914.* London, 1972.

Barnett, C. *Britain and her Army.* New York, N.Y., 1970.

Chegwidden, I. S. "The Imperial Defence College." *Public Administration* 25 (1947): 38–41.

Chisholm, C. *Sir John French.* London, 1915.

Fuller, J. F. C. *Memoirs of an Unconventional Soldier.* London, 1936.

Godwin-Austen, A. R. *The Staff and the Staff College.* London, 1927.

Hartley, F. R. "Science Graduates in the Army—Luxury or Essential?" *Journal of the Royal United Services Institute* 103 (Dec. 1978): 53–56.

Luvaas, J. *The Education of an Army, British Military Thought 1815–1940.* London, 1965.

Marder, A. J. *From the Dreadnought to Scapa Flow.* London, 1961.

Parker, P. *The Old Lie—The Great War and the Public School Ethos.* London, 1987.

Terraine, J. *Douglas Haig, the Educated Soldier.* London, 1964.

Veith, C. " 'Play Up! Play Up! and Win the War!' Football, the Nation and the First World War, 1914–1915." *JCH* 20 (1985): 363–78.

Young, F. W. *The Story of the Staff College 1858–1958.* Camberley, 1958.

Modern USSR

Cross, R. "Significant Parts of Kulikov's Study of the Soviet General Staff Academy." Combat Research Center, College Station, Tex., 1982.

Danchencko, A. M., and I. F. Vydrin, eds. *Military Pedagogy, a Soviet View.* Moscow, 1973 (USAF translation, Washington, D.C., n.d.).

Druzhinin, V. *Concept, Algorithm, Decision.* Moscow, 1973.

Erickson, J. *The Soviet High Command.* Boulder, Colo., 1984.

———. "Soviet Military Manpower Policies." *AF&S* 7 (1980–81): 29–47.

Head, R. G. "Soviet Military Education." *Air University Review* 30 (Nov.–Dec. 1978): 51–57.

Jones, E. *Red Army and Society, a Sociology of the Soviet Military.* Boston, Mass., 1985.

Keep, J. *Soldiers of the Tsar. Army and Society in Russia 1462–1874.* Oxford, England, 1985.

Kuhlov, V. *The General Staff Academy.* Moscow, 1976.

Lomov, N. A. *The Military-Scientific Revolution.* Moscow, 1973.

Miller, A. *Dmitri Miliutin and the Reform Era in Russia.* Charlotte, Va., 1968.

Odom, W. "Bolshevik Ideas on the Role of the Military in Modernization." *AF&S* 3 (Fall 1976): 103–20.

Seaton, A. *The Russo-German War.* London, 1971.

Seaton, A., and J. Seaton. *The Soviet Army.* New York, 1986.

Scott, H. F., and W. F. Scott. *The Armed Forces of the USSR.* Boulder, Colo., 1981.

Modern USA

Ambrose, S. E. *Duty, Honor, Country: a History of West Point.* Baltimore, Md., 1966.

Ball, H. P. *Of Responsible Command.* Carlisle Barracks, Pa., 1983.

Bowen, H. R., and J. H. Schuster. *American Professors.* New York, 1986.

Brewer, T. L. "The Impact of Advanced Education on American Military Officers." *AF&S* 2 (1975–76): 63–80.

Crackel, J. "The Founding of West Point—Jefferson and the Politics of Security." *AF&S* 7 (1980–81): 529–44.

Cunliffe, M. The American Military Tradition." In H. C. Allen and C. P. Hill, eds., *British Essays in American History.* New York, 1957, pp. 207–44.

Dastrupp, B. L. *The US Army Command and General Staff College—A Centennial History.* Manhattan, Kans., 1982.

Ellis, H., and R. Moore. *School for Soldiers: West Point and the Profession of Arms.* New York, 1974.

Euliss, J. P. "War Gaming at the U.S. Naval War College," *Naval Forces* 5 (1985): 96–106.

Forman, S. *West Point: A History of USMA.* New York, 1950.

Harris, S. E. *A Statistical Portrait of Higher Education.* New York, 1972.

Hart, G., and B. Lind. *America Can Win.* Bethesda, Md., 1986.

Hartmann, F. H. "The Naval War College in Transition." *Naval War College Review* (1984): 6–13.

Hattendorf, J. B., et al. *Sailors and Scholars.* Newport, R.I., 1984.

Hong, D. S. "Retired US Military Elites." *AF&S* 5 (1978–79): 451–66.

Houle, C. O. *Continuing Learning in the Professions.* San Francisco, 1980.

Huntington, S. P. *The Man on Horseback.* New York, 1957.

Janowitz, M. *The Professional Soldier.* New York, 1960.

Jordan, A. A., and W. J. Taylor. "The Military Man in Academia." *The Annals of the American Academy of Political and Social Science* 406 (March 1973): 129–45.

Karsten, P. *The Naval Aristocracy: The Golden Age of Annapolis and the Emergence of Modern American Navalism.* New York, 1972.

Korb, L. I., ed. *The System for Educating Military Officers in the US.* Pittsburgh, 1976.

Luttwak, E. *The Pentagon and the Art of War.* New York, 1985.

Lyons, G. M., and L. Morton. *Schools for Strategy.* New York, 1965.

Lyons, G. M., and J. W. Masland. *Education and Military Leadership: A Study of the ROTC.* Princeton, N.J., 1959.

McGregor, E. W. "The Leavenworth Story," *Military Review* (May 1956): 70–71.

Margiotta, F. D. "A Military Elite in Transition, Air Force Leaders in the 1980's." *AF&S* 2 (1975–76): 155–84.

Masland, J. W., and L. I. Radway. *Soldiers and Scholars, Military Education and National Policy.* Princeton, N.J., 1957.

Menninger, W. C. *Psychiatry in a Troubled World.* New York, 1948.

Military Review (May 1956): entire issue.

Millett, A. R. "Military Professionalism and Officership in America." Mershon Center Briefing Paper, No. 2, Columbus, Ohio, 1977.

Murray, W. "Grading the War Colleges." *The National Interest* (Winter 1986–87): 12–19.

Pappas, G. S. *Prudens Futuri: The US Army War College, 1901–1967.* Carlisle Barracks, Pa., n.d.

Pickett, W. P. "Eisenhower as a Student of Clausewitz." *Military Review* (July 1985): 21–27.

Pusey, N. M. *American Higher Education 1945–1970.* Cambridge, Mass., 1978.

Pratt, W. V. *The United States Naval War College—A Staff Study,* app. C. Newport, R.I., 1954.

"Report and Recommendations of a Board Appointed by the Bureau of Navigation Regarding the Instruction and Training of Line Officers," *United States Naval Institute Proceedings* (Aug. 1920).

Report of War Department Military Education Board on Educational System for Officers of the Army (Gerow Report). Washington D.C., April 1946.

Sarkesian, S. C., and W. Taylor. "The Case for Civilian Graduate Education for Professional Officers." *AF&S* 1 (1974–75): 251–62.

Shelburn, J. C. *Education in the Armed Forces.* New York, 1965.

Siddle, W. "The College Role in the Army School System." *Military Review* (May 1956): 5–14.

Taylor, W. J., and D. P. Bletz. "A Case for Officer Graduate Education." *Journal of Political and Military Sociology* 2 (1974): 251–73.

Upton, E. *The Military Policy of the United States.* Washington, D.C., 1901.

Vlahos, M. "Wargaming, an Enforcer of Strategic Realism." *Naval War College Review* (March–April 1980): 34–42.

Modern Israel

Dayan, M. *Diary of the Sinai Campaign.* London, 1965.

Gal, R. *Portrait of the Israeli Soldier.* Westport, Conn., 1986.

Lester, J. T. "Israeli Military Psychology." ONR Report R-13-73. London, Office of Naval Research, 1973.

Luttwak, E. N., and D. Horowitz. *The Israeli Army.* London, 1975.

Index